P. Louis Appia, And others

The Ambulance Surgeon or Practical Observations on Gunshot

Wounds

P. Louis Appia, And others

The Ambulance Surgeon or Practical Observations on Gunshot Wounds

ISBN/EAN: 9783337191825

Printed in Europe, USA, Canada, Australia, Japan

Cover: Foto ©berggeist007 / pixelio.de

More available books at **www.hansebooks.com**

THE

AMBULANCE SURGEON

OR

PRACTICAL OBSERVATIONS ON GUNSHOT WOUNDS.

By P. L. APPIA, M.D.,

FELLOW OF THE ROYAL SOCIETY OF NAPLES, OF THE MEDICAL SOCIETIES
OF GENEVA, FRANCFORT, NANCY, NEUFCHATEL, ETC. ETC.

EDITED BY

T. W. NUNN,

ASSISTANT SURGEON TO THE MIDDLESEX HOSPITAL,

AND

A. M. EDWARDS, F.R.S.E.,

LECTURER ON SURGERY IN THE EDINBURGH MEDICAL SCHOOL.

EDINBURGH:
ADAM AND CHARLES BLACK.
1862.

PRINTED BY R. AND R. CLARK, EDINBURGH.

PREFACE.

———◆———

In offering this edition of Dr. Appia's "Ambulance Surgeon" to the profession, the editors have only to state that the translation is slightly condensed, and that a few details to be found elsewhere have been omitted.

They have added a chapter upon Disinfectants. It was originally intended to add a series of illustrations of Surgical Appliances. But being unwilling to make the book larger than absolutely necessary, they have for the present limited the last part to the dressing of wounds, hemorrhage, and some of the more common varieties of fracture.

TABLE OF CONTENTS.

PART FIRST.

GENERAL INTRODUCTION — GUN-SHOT WOUNDS.

CHAPTER I.

CHAPTER II.

CHAPTER III.

PART SECOND.

GUN-SHOT WOUNDS CONSIDERED IN DIFFERENT PARTS OF THE BODY.

CHAPTER I.

CHAPTER II.

FRACTURES OF THE THIGH—

CHAPTER VIII.

CHAPTER IX.

PART THIRD.

SURGICAL APPLIANCES.

ERRATUM.

Page 54, line 27, *for* tire-bowrie, *read* tire-Boulet.

PART THE FIRST.

———◆———

CHAPTER I.

1. Old Delusions.

At the present day people no longer believe that gun-shot injuries possess the characters of burns or poisoned wounds. But that idea was dominant till the days of Ambrose Parée, and most irrational means were used to promote the separation of the eschars, and to neutralise a poison supposed to be developed by the bullet.

Thence the custom extended to that period when cauteries at a white heat were prepared to destroy the infected tissues, and deep incisions were made to facilitate the elimination of the poison, for which boiling oil of turpentine was considered an antidote.

These wholly theoretical views on the specific nature of gun-shot wounds are similar to their belief in former ages, that there were specific remedies for wounds of the different organs ; thus spermaceti, it was thought, hastened catricization after wounds of the lungs, and white of eggs or chicken's blood cured those of the eye.

We only mention these specimens of the errors of

an earlier age as historical curiosities, and to give us an opportunity of explaining the origin of some opinions which we may still hear advanced from time to time, such as the necessity for laying open a wound as a precautionary measure, under the impression that it changes the gun-shot wound into one of a simple character.

Perhaps there may still be some who believe in the *wind of a ball* (Luft streifschüsse ; wind contusions, *vent de boulet*).

It had been observed, as is well known, that heavy projectiles, especially balls, could produce deep and serious injuries to the soft parts, and even to bones, without necessarily breaking the skin, and these undeniable facts were accounted for by the extreme pressure which the air in front of the projectile underwent. But one need not be a natural philosopher to see that air is too delicate and elastic a medium not to separate on either side of a convex and limited surface like that of a bullet, rather than undergo extreme compression from it.

Besides, how often has one seen in battles, missiles carry away a piece of the uniform, the shako, nay, the very hair of a soldier, without hurting him. Thus, on one occasion, the tip of a nose was taken off by a bullet ; on another, the auricle, and without any further accident. At the bombardment of Antwerp in 1814, a ball shattered the legs of two officers who were each holding by the arm a third, placed so near them that the ball must have, it is conjectured, passed between his legs. Again, how could the compressed air alone cause these lesions, since the column of air is in front of the projectile, which, being necessarily close behind, must also reach the body.

Internal injuries formerly attributed to the pressure of the air (or wind of the ball) are evidently due to the obliquity with which the bullet impinges on the body and the diminution of its force at that moment.

A bullet which strikes the body obliquely is repelled by it, and the more it has lost of its velocity the more readily does this occur. These two circumstances combined, while they explain how the skin remains uninjured, do not prevent the occurrence of serious internal injuries.

The bones, in consequence of their rigidity, do not yield, like their soft coverings, to the pressure of a missile which retains a degree of velocity sufficient to cause comminuted fractures ; these again entail internal lacerations which may not extend so far as the integumental covering.

Larrey cites several cases where the ball produced serious injuries, sometimes mortal, without injuring the skin ; in these cases, the projectile had struck the body towards the end of its course, but so feebly, that the wounded person could give an account of what had happened some moments after. Here it would have been easy to have obtained confirmation of the *vent de boulet* theory.

The Crimean campaign afforded many instances of internal lacerations with unbroken skin, which in former times would have been attributed to the wind of the shot.

Dr. Quesnoy saw an engineer officer at the head quarters ambulance who had his forearm broken, but without any external symptom of injury. At Alma they took into the ambulance a soldier whose forearm was in its interior a mere mass of pulp, though his skin was unhurt. There were similar effects from the howitzer shell which wounded General Canrobert in the same battle ; his skin was hardly touched, but a part of the pectoral muscle was ground down (*broyé*). It was in this way General Mayran was killed, but in his case the injury was deeper seated, and although the skin was not torn, muscles, two ribs, and the lung, were crushed.

2. Shock.

Immediately a ball strikes the human body, general symptoms appear which are difficult to describe, but which are familiar to every army surgeon. All the bodily economy seems shaken by the severe injury, the immediate effects of which are as yet very limited. The patient is first seized with a convulsive trembling of his whole body, resembling the rigor which precedes a fever ; then come faintness, convulsions, general coldness of surface, a small, or perhaps scarcely perceptible pulse, and a livid countenance, bearing witness to the extreme nervous depression.

The wounded of the French Revolution, June 1848,

who were treated in the hospitals, were not always taken there immediately, so that we had not always the opportunity of observing this condition. But it was easy to satisfy oneself when all was passed that many of them had exhibited symptoms even to that of approaching dissolution ; but in twenty-four hours all immediate danger seemed past, and the patient was undergoing the usual treatment for one in his condition. This state of shock, apparently so serious, is nevertheless not always indicative of a very serious injury ; it may be no more than the reflexion of great nervous susceptibility.

For, although in penetrating wounds of the chest, abdomen, or cranium, it aggravates all the alarming symptoms immediately proceeding from the injury, it may also be associated with a simple wound of the soft parts. However, the general shock of which we are about to speak, is not *a constant* symptom.

In the Crimea, central disturbance, local or general insensibility, nay, even pain, were not present in some cases. Men with their upper and lower jaws crushed by fragments of shells, were known to walk from the trench to the ambulance. One of these men, from whose pharynx some fragments of bone were removed, though unable to speak, could write what he wished with a steady hand. The majority preserved their mental faculties unaffected by their frightful mutilations.

At Alma we have seen men whose limbs hung to them by a mere shred of skin, in the full enjoyment of their senses. General insensibility, which was rare,

appeared rather to accompany very extensive mutilations. Yet Guthrie tells of having seen a young sailor
in London who had had his entire upper extremity
carried away by a round ball fired from one of the forts
at Guadaloupe in March 1808. His body felt no
special sensation, and his senses were noways embarrassed.

Different explanations have been sought for this
sudden exhibition (explosion) of alarming symptoms,
which would appear to be effects almost peculiar to
gun-shot wounds. It seems to us that they ought not
to be solely referred to the nature of the weapon which
inflicts them, but that they are in a great measure due
to the over-excited condition of the subject.

Most of those wounded are, when struck, in a state
of agitation and excitement, which is sufficient to
account for the symptoms we have described.

It is a mistake, then, so far as we can see, to account
for this shock to the whole bodily economy merely by
the nature of the injury which has occurred. Were
such the case, the amount of shock should be in exact
proportion to the importance of the organ wounded, but
experience does not prove this to be the case. In
addition to the symptoms already described, the
wounded limb is sometimes seized with a numbness
approaching to insensibility ; this. symptom generally
soon passes away, but it may perchance be taken advantage of in those operations which one is called upon to
perform immediately. The stupor is, indeed, sometimes
so complete, that the wounded man can undergo, with

but little suffering, operations which in general are acutely painful.

Of course we are not alluding here to the anæsthesia resulting from the injury of a nerve. That complication is by no means rare, often lasting for a long time, and may continue for an indefinite period, even after the wound has been apparently cured. This condition is analogous to paralysis or coma from wounds of the encephalon.

M. Quesnoy has seen soldiers continue fighting, unconscious that they were wounded; others sustaining serious injuries involving bone, and yet who had merely felt a sensation like the stroke of a cane.

In general, pain is a late symptom of a gun-shot injury, coming on as inflammation is developed.

3. VARIETIES IN GUN-SHOT WOUNDS.

Wounds from fire-arms are of infinite variety, according to the velocity of the projectile, its bulk, shape, and direction with regard to the body, and also the numberless changes of posture the latter may assume at the very moment when the accident occurs.

It is especially this last very variable circumstance, which explains the extremes of difference between cases of gun-shot wounds and those compound injuries which seem due to totally different causes, though they are the result of one single projectile. Indeed, the state of health a wounded man may be in when struck, will influence, though in an extremely variable way, the

progress of the wound. In short, all these circumstances are so many factors which enter into the calculation of our pathological problem, varying the result
indefinitely.

But after all, these gun-shot injuries have this advantage over other wounds, that the circumstances
under which they occur are already known, and that
they possess a very uniform character as regards diagnosis and treatment.

There is always a rounded projectile penetrating into
the body with great velocity; so the different cases are
capable of explanation and a more rigorous comparison
with others than the injuries from all causes received
into a general hospital could possibly be. It is very
probable that before the use of gunpowder it must
have been still more difficult to lay down rules for
military surgery, as in those days injuries depended so
much on individual prowess.

As to the relative frequency with which the different parts of the body are struck by the bullet, I would
refer to the following table, borrowed from Dr.
Serrier :—

	Cases.		Cases.
Leg . .	100	Shoulder .	42
Thigh . .	97	Skull . .	37
Face . .	61	Forearm .	36
Arm . .	60	Knee joint .	54
Hand . .	57	Foot . .	29
Chest . .	53	Elbow joint .	22
Abdomen .	52	Neck . .	22

	Cases.			Cases.
Genitals	. 18	Hip	.	. 6
Ankle joint	. 15	Vertebræ		. 10
Shoulder	. 13	Wrist	.	. 2

Total . . 784

Much labour has been expended in endeavouring to find the distinctive characteristics between the two wounds which a bullet leaves in the skin.

That question only offers a very slight attraction to the practical surgeon, and has scarcely any useful application except in medical jurisprudence. Nevertheless, the majority of surgeons have discussed the matter at great length, and in a most controversial spirit. It seems to us that practical surgery has derived little benefit from these researches. So let us only linger for a short time over this subject, the more as our own observations on it are unfortunately somewhat incomplete.

It has been generally remarked that the orifice of entrance is smaller than of exit, its margins more sharply cut than the latter, which is usually swollen and with everted edges. In the Parisian hospitals, where, in 1848, hundreds of wounded were collected, I sought often to establish this difference, but I did not find it so well marked as has been generally described. It is true, however, that when I made up my list of cases several days had elapsed since the memorable battles in the faubourgs.

The skin surrounding the wounds had had time to

contract by its own elasticity. Suppuration was established on each side of the track of the bullet, so that
one could no longer distinguish the difference between
the wound of entry and of exit.

But, nevertheless, the difficulty we have so constantly found in establishing a difference between the
two apertures, shews that the current opinion is not
always well founded. Amussat even says that he has
found the wound of entry the larger of the two.

It is necessary, I believe, for the rational solution
of this question, not to presuppose that it must be
explained by any one sole and absolute rule, but that
different opinions will hold good under different circumstances under which the wounds are examined and the
parts injured.

M. Blandin shares the same opinion as Amussat,
and even considers that he was the first to demonstrate
that the wound of entry is the larger, except when the
ball has encountered a bone in its passage. What
makes the orifice of exit the smallest, he thinks, is the
elasticity of the skin.

M. Velpeau entertains both opinions, that is to say,
he has found the relative size of the aperture vary
according to the form of the missile and the kind of
tissue which it traversed. However, generally, the
wound of entry was the larger. M. Hugvier has made
many experiments to determine the manner in which
the bullet is affected by different physical conditions
and among different tissues. According to him, the
wound of entrance is in general accompanied with loss

of substance, while the other has shreds and swelling outside of the tissues carried away by the ball. M. Baudens says that the orifice of entry into living tissues is depressed, round, regular, and smaller than that of exit, which is swollen outside. According to the velocity of the projectile, the wound presents either

a. A simple bruise, without laceration of the skin;

b. A wound with a single orifice ;

c. A wound with a double orifice ;

d. When it has carried off a limb.

But of all the differences relative to projectiles the form of the bullet is most interesting to surgeons.

In 1848 we only saw spherical balls. Since then they have introduced cylindro-conical ones into the French army. They were largely employed in the Crimean war, and observations there proved that their effects differed remarkably from the old balls. The perfection to which fire-arms have attained has given to the cylindro-conical ball the greatest conceivable velocity and a rotatory movement on itself, which has most severe effects upon a bone. A terrible shock, splitting the bone in every way, has replaced the old injury done by the round bullet. This comminution of the bone has no parallel in former surgical annals. The great number of splinters, the disturbance to the entire bone, the longitudinal cracks—the necessary consequence of such a shock—exhibit a degree of seriousness proportionate to the injury inflicted.

The most serious consequences of wounds from conical bullets would seem to depend on three causes.

1. The conical ball is never turned by a hard or elastic body, but passes straight through it.

2. It may, nevertheless, in its course through the body, change its longitudinal position, so that it strikes organs with its long axis, and causes very considerable damage.

3. It is probable, from the pointed shape of the conical ball, that it causes less actual loss of substance, but, at the same time, more lateral separation of the tissue, from its wedge-like form.

The form of orifice, too, both of entry and exit, is different from those caused by the round ball. The entrance is a well-defined oblong, sometimes almost linear, so also is the exit, unless the ball has changed its direction with regard to the body, when the opening is large and lacerated.

Dr. Quesnoy thinks that, from all these peculiarities, the conical ball more frequently causes fractures, with great numbers of splinters.

So the surgical experiences of the Crimean war have been rather discouraging as regards the resources of art for preserving limbs which have sustained comminuted fractures. Hardly had improvements in transport and the dressing of the wounded given surgeons the hope of being able to preserve limbs which shortly before would have been doomed to amputation, when that hope is destroyed by the disastrous effect of the conical ball.

Thus the art of destroying life seems destined to advance with a speed which the science intended for its preservation is never able to attain.

4. Foreign Bodies.

No one is ignorant that one of the most ordinary complications which aggravate the progress of these wounds, is the presence of foreign bodies ; these are—
1. The destroyed tissues.
2. Bone splinters.
3. The ball itself.
4. Pieces of clothing, woollen, or such other objects encountered by the ball in its passage.

Destroyed tissues contained in the wound are not, properly speaking, foreign bodies. They have no other inconvenience than hindering a cure by first intention, a fortunate result which one very rarely observes, although John Hunter says that it is possible when the wound has only involved the soft parts.

The form in which a bone has been fractured varies much with its anatomical structure. The most numerous splinters are produced in the shafts of long bones, of which the consistence is hardish, and which contain scarcely any spongy tissue ; so is that of all fractures the most difficult to consolidate. Fractures of bones are sometimes accompanied by fissures or cracks which may extend far beyond the point injured by the ball. Guthrie, for instance, has often seen them in fractures of the middle of the femur extend to the condyles, and cause ulceration of the articular cartilages of the knee joint.

The pathology of gun-shot wounds would be of essential service in solving a number of the problems

connected with them.   Unfortunately that branch of
surgical science is as yet but little advanced, probably
owing to the scanty leisure available for such researches
by the army surgeon on actual service.   To explain
the mechanical effects of bullets upon bones, it would
be necessary to make a series of experiments on the
dead subject.   These, which so far as I know have
not yet been made, would be of more service than those
made on inorganic bodies.

The external injury being generally insignificant
compared to the internal destruction of parts, one is
naturally led, from external examination, to under-esti-
mate the mischief which has occurred within.   One is
apt, then, from this superficial view, to entertain a hope
of cure, which turns out detrimental to the chances of
our poor patient, and from which one is only warned
by repeated disappointments.

Hence it is that in our opinion there should be so
much more attention paid to the pathological anatomy
of gun-shot wounds than has hitherto been.

The velocity of the ball influences the extent of
its injuries to a bone, and it is generally thought that
these effects are less in inverse proportion to its velocity.
As we see holes made in panes of glass by the passage
of a bullet, the quicker the missile has passed through,
the cleaner will be the margins of the orifice, the less
the extent of the cracks.

If wounds of bone vary according to the velocity of
the ball, the latter in its turn is changed by its encounter with a hard obstacle to its course.   Thus in

one case, a ball which had pierced the chest near the external border of the left scapula, and which was removed at the autopsy in a mass of broken-down lung tissue near the rib, was flattened out and thinned like a piece of cloth.

In a second case, the ball, which had entered into the region of the great trochanter, was removed three months after, while the patient was yet alive, in a cavity of an enormous callus, quite flattened out. In a third, a ball which had smashed the acromio-clavicular fossa was found quite altered in shape.

M. Huguier has collected the most curious cases which he met with among the wounded of the Revolution, and arranged them thus—

1. A ball split on the edge of the petrous bone.
2. A ball split in two by the crest of the tibia, which broke the latter, and half remained in the periosteum.
3. A ball divided by the orbital arch into two parts, the larger of which lodged behind the eye, at the bottom of the orbit.
4. One split into three parts by the orbital arch.
5. Division into three parts by the edge of the clavicle.

Percy speaks of a ball shot into the skull of a subject, which spread out on the internal layer of the skull like a plate of tin. M. Baudens relates that in a femur he removed from an officer at the Sig bivouac, the bullet was found among the splinters divided into two exactly equal parts. In another soldier, a bullet which had struck the great trochanter was divided into three separate pieces.

But certainly the oddest example is the one related by M. Serrier. In Algeria a ball broke into five fragments on a rock five or six paces from a grenadier; the first fragment struck and broke his right ankle, two others pierced further down, the fourth wounded his right thigh, and the fifth lodged in the skin of the back of his head.

Pieces of clothing and of the uniform often accompany the ball in its course, and complicate the treatment of the wound. The presence of these bodies, however, is not generally a very serious matter. In general producing but little irritation, though often of long continuance, and one sometimes removes a piece of trouser or cape, etc., from a wound in which the suppuration has been reduced to a minimum. One sees, on the other hand, a delinquent body pass out of a wound, explaining why the latter, which in other respects had a healthy appearance, was so tardy in healing. We have in this way seen washed out with the pus some cloth of a cape from a wound in front of the sternum.

The following is a list of some of the foreign bodies found in thirty-one cases in the revolution of 1848.

Cases.
5. Little portions of ball.
2. Small shot.
3. Pieces of wadding.
3. Pieces of shoe.
6. Pieces of cloth and shirt.
4. Pieces of wadding and tow.
2. Pieces of worsted.

Cases.

1. Bundle of hair.
1. Many hog's bristles.
1. Piece of cast iron.
1. Small piece of wood.
1. Fragment of copper ornament from shako.
1. A nail.

A very curious case is related by Laroche (Surgical Narrative of Events at Lyons in 1835). One of his relations had twenty napoleons in his pocket, which, struck by the ball, were driven into his belly, and were all more or less spoilt. In the Crimea one has seen fragments of shell the size of two nuts, lodged in the abdominal parietes, and at other times in the thigh and leg.

It is not very rare when heavy projectiles, as bullets, bomb shells, or *biscaiens*, have been discharged, to find pieces of them in the wounded limb ; the history of surgery presents numerous examples of this.

In 1848 we saw but few specimens of wounds from large missiles, as they were directed less against men than material obstacles. It is only a bullet which can penetrate the tissues and remain concealed amongst them. One would be tempted to doubt the stories told of them did they not come from such authentic sources.

For instance, the case which Larrey relates minutely of an artilleryman who was struck by a ball on his right thigh. The femur was broken; as for the ball, it pierced the thickness of flesh, turned round the bone, and ended by dipping near the anus into the hollow of the thigh.

When he was brought to the ambulance, neither he nor his surgeons suspected the presence of a foreign body. The patient even was of opinion that the same ball had passed on and struck another bombardier. It was only when performing amputation that Larrey discovered a ball five pounds in weight.

Dupuytren relates that a ball of nine pounds weight was so completely concealed in a patient's thigh, that the surgeon did not at first discover its presence.

On the morrow after the taking of the Mamelon-Vert, a soldier applied at the ambulance, said to be wounded in his left thigh. About its middle was found a small circular aperture, like that from a round ball, not a wound of exit. On examination they could feel an obscure swelling in the popliteal space, but otherwise there was no swelling, redness, or especial amount of pain. A large incision enabled them to discover and extract an enormous shot which had run round the femur without breaking it.

From the external appearances, perhaps, of a hundred surgeons, fifty might have thought no ball had entered, but assuredly a hundred would have denied that there lay there a *biscaien*, but it was so, neverthe-less.

5. Twisting Course of the Ball.

We cannot pass silently one question which has been often under the consideration of surgeons, and has never been satisfactorily explained. I allude to the winding course of a bullet.

One frequently meets with two apertures, so placed with regard to each other, as to appear quite independent. Thus a ball which has entered the ankle joint goes out at the knee ; another, piercing the forehead, escapes at the temple. I have seen wounds of the genitals, where the ball entered above the glans, had got out through the left thigh without leaving any intermediate traces of its passage.

In other cases, two apertures opposite each other, including between them, in the straight line which must unite them, organs important to life, which, if the ball had touched, it must inevitably have been followed by death, whence one naturally concludes that the ball has passed round these organs. How are we to explain, except by some exceptional disposition, such a case as this from my own experience ? the front of the knee is struck by a ball which escapes at the back of the joint, leaving no more mark than an ordinary seton wound. So there are two cases of wound of the pelvis marked in our observations where the ball entered the left groin, escaped by the buttock without hurting any important organ, probably because the ball had glanced along the pelvic cavity. These will be found in our chapter on wounds of the pelvis. Another occurred in Roux' practice. A simple perforation of the right shoulder with no trace of fracture, but nevertheless a line drawn between the two apertures passed straight through the head of the humerus. Dr. Hennen de-

clared he saw a case in which the ball entered near the
thyroid cartilage, and which, after going round the
neck, returned to the same point by which it had
entered, and was extracted at that spot.   In another
instance, a soldier was struck at the moment when he
extended his arm to mount a ladder.   The ball entered
the middle of the humerus, passed along the limb above
the posterior aspect of the thorax, opened for itself a
passage in the abdominal muscles, pierced those of
the buttock, and passed again upwards to the anterior
aspect of the opposite thigh ; on another occasion, a
ball penetrating the thorax found its way to, and
lodged in the scrotum.   One of my colleagues remarked
after the battle of Novara several of these cases of
wandering balls (cas de contours de balle).   In one
case the ball seemed to have crossed the neck from
side to side.   Both openings lay right and left of the
larynx, but a subcutaneous ecchymosis joined these
two wounds, and shewed that the ball had passed round
the trachea.   The patient made a rapid recovery.

I know not whether this turning of bullets, which
I do not question by any means, has been proved by
pathological anatomy.   It would be very desirable to
have the question determined by positive proofs.

I have sought in vain, in many cases under my
observation, in hospitals, to satisfy myself of the wandering of the ball by an examination of the wound ;
nothing has indicated it, and I have not been able with
others to verify the external redness, said to be one of
the ordinary signs.   No more is the spitting of blood

a pathognomonic symptom of a penetrating wound of the lung, for one has seen simple contusions and superficial wounds complicated by it.

It is also rare for the turning of a ball to be proved by post-mortem examination, probably because offering a favourable chance for a cure ; as a general rule, an autopsy was not required. (Refer on this point to the third observation on purulent absorption, cap. iii.)

One is forced to admit very often that there has been a deviation in the course of the ball in cases when the patient's progress has been too favourable to allow us to believe that the ball has traversed any vital organ, and so to suppose that it penetrated in a direct line, when the severity of the symptoms seems more in proportion to the importance of the organs injured.

CHAPTER II.

On Diagnosis.

THE diagnosis of gun-shot wounds is not always easy to determine. There are other difficulties besides the turning of the bullet, which we have just remarked.

It is generally necessary, in order to determine the treatment of a wound, to know its depth and direction. As we have just seen, the latter is not always in a direct line between the two apertures.

Sometimes we cannot draw such a line, and in consequence, we must explain the peculiarity by assuming that the wounded man was in a particular position, which sometimes he remembers and tells us. At other times, the course presents three or four openings, and it is then enough to line them out to find the position the body was in when struck. The most frequent examples of this description are changes in the relative positions of the arm to the trunk. The arm being often extended and put into play, whether to attack with the bayonet or the sabre, a ball when it arrives passes successively through the arm and the chest in a direction which can only be explained by imagining the person for a moment to resume the posture he was in when wounded.

In one case which I saw at St. Louis, the ball had traversed the left biceps muscle, then had penetrated the chest by the axilla, and had gone out again above the left lumbar region ; he went on well. To understand the course of this ball, one must imagine the body much bent forward, and the left arm extended to the uttermost.

Second case. Long wound of the thorax, the ball entering near the right false ribs and going out in the neighbourhood of the spleen. At the same time, the ball went through the right arm, which it paralysed, probably from division of the median nerve. Here, too, we can only suppose that the arm was extended.

Third case. The ball entered in the upper third of the right arm, and went out just over the nipple. If the arm is hanging, the straight lines uniting these two wounds to the body, would, in a manner, seem to indicate a necessity for *four* skin wounds. But as there were only *two*, we must imagine that the arm was stretched out when struck.

How can a ball, which has penetrated the root of the nose and pierced the velum palati, get out by the mouth without extending further its disastrous effects ? The case to which I allude I saw in 1848. You could pass your little finger through a great opening in his palate ; there was loss of bone and fracture of some teeth. But further than that he did not suffer, spoke without pain, and much as one does after losing some of his palate from syphilis.

The inferior extremities being, during action, less frequently approximated to the trunk, never present the same complications as the arms do, as we have just described. Besides, their greater thickness impedes the ball, and prevents its injuring other parts of the body. Indeed, we meet with but few cases where the ball has broken both thighs, or even both legs, at once ; nor others where a wound of the thigh and of the trunk have been due to one single projectile.

We possess, however, two examples where the same ball has wounded both limbs. On the 18th June 1848, a man received a ball in the right arm above the elbow, producing a comminuted fracture. On the left side the same ball entered below the elbow, and fractured the upper part of the radius. Death took place from pyæmia.

In another case the ball traversed the soft parts of one thigh and glanced off the other. The most common direction of wounds of the lower extremities is from before backwards, involving a variable thickness of the limb.

When in the chest, the ball's direction will be sometimes, indeed generally, antero-posterior, at others parallel with the lateral diameter. It is rare to meet with a case where treatment is required when the ball traverses the middle of the thorax, as death is generally instantaneous. More frequently, the courses are superficial, or even simple perforations only involving skin. The direction of wounds of the head and abdomen is easy to determine by examination, but even in these regions we must refer occasionally to the deviations of the ball to explain unexpected and, if one may so say, unphysiological recoveries. As an instance, I select from my notes a case in which the ball entered below the umbilicus and passed out at an opposite point in the lumbar region. Though it had thus traversed the abdomen from side to side there was no symptom of wounded intestine, no tympanitis, no pain, but a complete recovery.

As to the *depth* of a wound of the splanchnic cavities it is useless and wrong even to have an inclination to determine its depth by the probe. This practice, which I have seen some surgeons delight in, apparently to enhance the apparent importance of their own functions, should be especially repudiated. The treatment will be the same whatever the depth of the wound.

All authors agree on this point, so it may be considered settled. In the Crimean campaign both the French and English surgeons were of this opinion. For the limbs, on the other hand, it is often important to satisfy one's self of the nature and extent of a fracture; the probe may, in these cases, become an invaluable guide, and enable us to ascertain the presence of splinters, which we should remove, as well as the suitable after-treatment. It is from this examination, among other things, that we have to determine the grave question of primary amputation.

And we may add, that as this examination of wounded limbs is so necessary, it should be made at once, especially if the bone be injured.

The introduction of the finger, and especially of a probe, is always a painful operation, so it is well to perform it when the limb is still numbed by the shock of injury. No sooner is inflammatory reaction established than sensibility is excited in the same proportion, and the slightest touch produces intolerable suffering.

In many cases the finger is the best probe—the little finger especially, if of smaller circumference than the common bullet (16.7 m.), and its sense of touch detects what would be passed over by a metal instrument. This is a fact of almost universal application, and applies to other as well as gunshot wounds.

It all comes to this, the surgeon should use the greatest circumspection in the examination of penetrating wounds of the abdomen and thorax, and in their diagnosis, from what their general appearance presents to his consideration; but in wounds of the limbs he must explore them to the bottom, for on this will depend their treatment, and especially the propriety of amputation.

CHAPTER III.

PROGNOSIS AND COMPLICATIONS.

1. *Difficulty of Establishing a General Prognosis.*

IT is impossible to lay down general rules for prognosis in gun-shot injuries. Their importance depends less on the projectile than on the organ which is injured by it. There are therefore as many prognoses as there are possible organic injuries.

It is necessary, also, if one desires to arrive at a conclusion of value, to ascertain the time at which the prognosis is made.

The mortality, it is thought, is greater in the first few hours. Professor Roux lost, in l'Hotel Dieu, in the first twenty-four hours, 25 out of 179 wounded ; at a later date about 35 ; forming, in all, a mortality of 60 out of 179—that is, a third. M. Giraldès, clinical surgeon to the faculty, gives, out of 47 wounded, 21 deaths soon after admission. At the Maison de Santé (faubourg St. Denis) they lost about a fourth, viz., 20 deaths out of 84 wounded. M. Baudens gives a sixth, or 28 deaths out of 164 wounded. M. Huguier, a seventh, or 22 out of 151.

These differences between authors is partly explained by the fact, that the wounded have come into their hospitals at different distances of time from the receipt of their injuries. Another cause, also, ought to have some influence on the results, viz., the time which has elapsed between the receipt of the injury and the date at which the estimate was made, for it is clear, the longer the time the larger will be the list of deaths. Again, the results obtained by M. Valette at Constantinople, who only received his wounded after five or six days, and the observations collected by other surgeons, must necessarily give a very different average of mortality. M. Valette is said to have only lost 10 out of 280, only one twenty-eighth, which is four times more favourable than the results in the Parisian Hospitals. But surely it would be well to know when these cases reached M. Valette, and what had been their average number of deaths before reaching Constantinople. Here is the list in M. Scrive's work on the Crimea.

Of those Wounded by Fire-arms in the French Army, who were admitted into Hospital in—

	Wounded.	Deaths.
October, 1854 . .	370	45
November „ . .	1210	196
December „ . .	550	82
January, 1855 . .	628	40
February „ . .	730	58
March „ . .	1484	127

		Wounded.	Deaths.
April, 1855	. .	1801	206
May „	. .	2888	333
June „	. .	6062	433
July „	. .	2058	288
August „	. .	3855	386
September „	. .	8665	1300
October „	. .	686	172
November „	. .	287	53
		31,274	3719

which gives an average mortality of rather less ; that is to say, 1 in 8½.

That is a proportion closely approaching the Parisian returns, and shews that the circumstances in which M. Valette was placed in relation to his patients were exceptional, and do not allow of comparison with those which embrace the treatment of cases from their very commencement. From all the lists we have referred to, it would seem that the average mortality from gun-shot wounds oscillates between a third and an eighth of the whole. But it would be necessary to have a list comprising many thousand cases to obtain a satisfactorily accurate general result.

Our impressions acquired in the earlier days spent in the different hospitals would have led us in general to augur favourably of the majority of the wounded ; and were we limited to this superficial opinion, we should have said that gun-shot wounds were less dangerous than the imagination represents them. But in the days which elapsed between the affairs of June and our first visits, the wounded had been decimated. Then we must remember that penetrating wounds of the chest can go on very favourably for several days, and then suddenly terminate fatally from rapid effusion.

We must also remember that complicated wounds of the thigh, although well put up, and presenting for several days a satisfactory appearance, almost invariably terminate fatally. It is necessary to guard against consecutive hemorrhage often breaking out at the moment when everything appeared to be going smoothly on to a wish, as well as against purulent absorption, which sometimes complicates the simplest wounds ; as, for instance, the seton in the neck I mentioned, which suddenly ended in a general infection that proved fatal. It is my opinion that there are in fact scarcely any, except wounds through the chest and those of the pelvis, which really present a more favourable prognosis than the importance of the wounded organs would seem to bear.

SURGICAL PROGNOSIS.

To appreciate correctly the prognosis of a bullet wound, it is necessary to have a considerable list of wounds of the same kind, with a description of the other general circumstances which may have influenced the result. Admitting the state of health to be pretty much of an average among the wounded, the prognosis of wounds becomes a physiological as much as a pathological question. All the organs may be wounded by the bullet, and all at different points ; the question then naturally arises—which are the organs that, being hit, necessarily involve death? The bullet then becomes what may be called the instrument of physiological experiment.

Wounds of the Heart, of the Lung, of the Brain, the tripod on which life rests, will be generally fatal when they reach the centre of the organ, as the base of the brain, the root of the lung. As to the heart, a wound of it seems incompatible with life. However, M. Jobert cites the case where a bullet remained three years in a heart without producing suppuration.

Cures of the *parenchymatous* parts of the lung are, as we shall see, common enough. Wounds of the brain, being necessarily accompanied by fracture of the skull, are aggravated by that complication. They generally terminate fatally, as will be seen hereafter.

The *Spinal Marrow* cannot be wounded without causing death, whether from its importance to life, or from the extensive osseous injuries which of necessity accompany it.

Penetrating Wounds of the Abdomen are almost always fatal, owing to the impossibility of retaining the edge of the wounded intestine in a suitable position for cicatrization. Nevertheless, we shall see that a ball can traverse the belly without causing death.

Wounds of the Liver can recover with an hepatic fistula. We possess two instances.

Lacerations of the Bladder are almost always fatal,

from urinary infiltration; but there are nevertheless
some brilliant exceptions in the history of surgery.
As to—

Fractures of Bone, their prognosis depends on
several causes, and first, upon the degree of splintering,
necessarily involving the cohesion of the bone and its
vitality more or less, that is, the rapidity with which
new bone will be thrown out.   This rapidity of reproduction, in its turn, depends on the part of the bone
broken, as well as the health and the age of the patient.
The danger which, under these circumstances, fills the
surgeon's mind with anxiety, is less the organic obstacles to the reunion of the bone, than the presence of
a permanent centre of suppuration, liable at any moment
to lead to some dangerous complications.

Fractures of the Skull owe their unfavourable prognosis, independently of the extent of the injury, to the
inflammation which they set up, often slowly and insi-
diously, from without inwards, through the thickness
of the cranium to the cerebral mass.

A ball may lodge in a vertebra without causing
death.   In one case which we saw, the ball had entered
the right external angle of the sacrum, a condition
favourable enough.   In another case it entered the left
side of the chest, and was taken out from one of the
lumbar vertebræ.

Wounds of the Pelvis admit a much more favourable
prognosis than fractures of the long bones.   Perhaps
the ball can more readily run round the convex or concave surface of the pelvic bones, without injuring important organs.

With regard to the Limbs, we shall see, that of all these wounds, the most serious are those of the femur in its whole length, of the knee, and of the hip-joint. Complicated wounds of the leg and upper extremity are often cured ; those of the foot, as, in general, those seated in an inelastic organ, seem predisposed to produce peculiar nervous symptoms, especially tetanus.

The history of military surgery is full of anecdotes of cures obtained in the face of chances which seemed to be necessarily mortal. It is difficult, in many cases, to say why an apparently trifling wound terminates fatally, why another has run the gauntlet of the most calamitous circumstances, and been cured, in some way or other, in spite of our prophecies. Thus, at the charge at Eupatoria, one dragoon received seventeen lance wounds—they thought him a dead man. At this day he is one of the dashing troopers of the 3d French huzzars.

The most interesting form of statistics upon the effects of gun-shot wounds, would be afforded by a visit to the Invalides. This fine institution contained, in *April* 1851, 3167 invalids, who are here arranged according to their rank—

2 Lieutenant-Colonels.

1 Chef de Bataillon.

72 Captains.

260 Lieutenants.

933 Other Ranks.

1899 Privates.

Of these, 1481, or rather less than half, had no wounds.

Of the remainder, there were the following, who had
been subjects of gun-shot wounds —

Disarticulations at the shoulder joint—

„	„	„	right	.	4
„	„	„	left	.	11

Amputations of both arms . . . 1
Excision of head of humerus—left . . 3

„	„	„	right	.	1

Amputations of—

Right arm 59
Left do. 48
Right forearm (one also lost a leg) . 24
Left do. 26
Wrist 12
Hip-joint 1
Right thigh 60
Left do. 42
At knee-joint 1
Both legs 8
One do. 165
Partial of foot . . . 43
Blind by balls 10
 „ by powder 10
One eye lost by ball . . . 20
Wounds of—

Skull 88
Face 143
Neck and nape . . . 28
Back and thorax . . . 69
Abdomen and loins . . . 20

Wounds of—

Pelvis, buttocks, and perineum	. 46
Genito-urinary organs . .	. 16
Shoulders and arm-pits	. 79
Arms	. 117
Fore-arms . . .	. 156
Elbow	. 74
Hip-joint . . .	. 25
Thigh	. 233
Knee and ham . . .	. 73
Legs . . .	. 330
Feet	. 121

We shall consider, further on, the relative details of complicated wounds of the limbs, involved in these lists, when discussing the relative merits of amputation of the thigh, and attempts at preservation.

2. MEDICAL PROGNOSIS AND GENERAL COMPLICATIONS.

After the relative importance of the organ wounded, the prognosis is immensely modified by the general health of the sufferer, and the hygienic conditions in which he is placed. How immensely does the diseased habit of body influence the progress of a wound. If an atmosphere of typhus, cholera, dysentery, decimates lives which the bullet has spared, must they not still more compromise the progress of the wounded. Are not the putrefaction, the necrosis of hospitals, merely the local reflections of a generally infected state due to wholly different causes than the nature, form, and situation of the wound.

Every surgeon has seen the most simple wound converted into a mortal injury by some simple, general cause, and the most complicated injuries get well owing to a good constitution and well-managed hygienic arrangements.   It is so important to bear in mind the epidemic influences, that a practitioner who could only take into consideration the surgical circumstances of the wound, and calculate his patient's chances by them alone, would constantly be deceived in even his most carefully formed opinions.

One of the complications which often aggravate the condition of wounds, is *intermittent fever* (fièvre d'accès). Fever, as is known, destroys more soldiers than the bullet, and the fièvre d'accès is one of the most widely spread.   Thus we see it stated in the statistics, that the regiments which suffered most in the Crimea were those which had previously gone through the campaign in Bulgaria.   The sojourn at Varna and the valleys of the Danube, was one of the most terrible causes of destruction to the Crimean army.

Therefore, what good reason there is to admire men who displayed such constancy and heroism, though already a prey to the most terrible elements of physical and moral annihilation !

A curious fact, worthy of remark, is the general and powerful effect *typhoid* fever has on the health of armies, that even where it does not shew itself openly, it seems, nevertheless, to be able to exist latent in soldiers who are apparently well. In the Crimea there were found ulcers in the intestines of men killed by the enemy's fire, in the midst of, to all appearance, perfect health. Intestinal ulceration thus seems the anatomical analogy to debility, offering a ground ready prepared for the influences of all kinds of poisons which hover over crowds of men placed in unfavourable sanitary conditions.

John Bell, in his treatise on wounds, insists on the bad character they may, especially stumps, suddenly assume.

This change, which sometimes occurs after the wound has gone on prosperously for several days, generally is produced, says Bell, under the influence of some constitutional derangement.

It is sometimes an attack of fever, the malignant dysentery, typhus, or sometimes simple diarrhœa. The wound then reflects externally the changes occurring within.

He recommends in these cases wine, sulphuric acid, and especially quinine.

In the Crimea, the wounds among the French usually exhibited a pale tint and a certain sluggishness in healing, circumstances dependent on general organic debility. The Russians in this respect shewed an unquestionable superiority to the French. In Paris, in 1848, on the contrary, the wounds were usually of a bright red, and the process of granulation active. We only observed in the whole of the hospitals two or three cases of hospital gangrene.

If sanitary conditions have immense influence, on the other hand, when we compare the wounded of the Crimea with those of the Revolution, the difference, especially in the serious cases, is not so sensible as one would be inclined to think.

If the wounded of Paris had been kept almost entirely sheltered from hospital gangrene, from *ostéomyélite*, and the acute diseases of a camp, they were, on the other hand, decimated by purulent absorption. We have often seen a patient carried off by that serious complication at the moment when his wound seemed advancing to perfect cicatrization ; and nevertheless *he* was in the most favourable hygienic circumstances, save a certain amount of crowding not to be avoided in a medical institution.

Of 19 wounds of the arm, marked in my notes, 7 submitted to amputation, and of these 4 sunk under purulent absorption. "We have had," says M. Roux, "a good deal of purulent absorption ; it is that which has in a great degree decimated our patients."

As an instance of the seriousness which this complication will suddenly give to a wound otherwise simple and of good aspect, I shall cite the following :—

Case 1. Simple penetration of the right arm without fracture.   In spite of the trifling nature of the injury, symptoms of purulent fever were developed.

At first a severe rigor, than a yellowish tint over the whole body, quick pulse, hot skin, dry rough tongue (saburrhale), tympanitis resisting pressure, general prostration, dorsal decubitus, sunken eyes, vacant countenance, suppuration of a moderately healthy character.

Some days later, suppuration almost arrested, pressure upon both orifices does not emit a drop. General state very bad, high fever. Six days later—death.

I come with some unwillingness to this case of a simple neck-wound, which was fatal after showing all of a sudden similar symptoms. I merely allude to this case to avoid repetition.

Case 2 serves also for an example of the twisting course of a bullet. A guard had received a ball in the thorax ; it entered about the middle of the collarbone, breaking it ; then it went out behind the infra spinous fossa, breaking the scapula. There was no injury to the side of the chest. No bleeding, no paralysis of the left arm.

The patient did well for some days, then, about the eighth, he had rigors, fever, and died at five in the afternoon.

The autopsy shewed that the ball had not pierced the thorax, it had traversed the axillary fossa, in a line with the costal wall, then the serratus magnus and subscapular muscles, and broke the scapula ; it passed exactly below the plexus of vessels and nerves, which were found in the midst of a blackish mass of tissue, among which it was difficult to recognise any particular organ.

In general, *the approach of symptoms of infection* began unexpectedly by a shivering fit, followed by diaphoresis, and then the general symptoms were developed, as described.

The progress of this formidable malady was in general very rapid, sometimes only lasting a few hours; its termination almost invariably fatal. Often it caused death before it was possible to show its approach by further symptoms than the chills and sweats.

Its arrival generally coincides in a sudden and alarming way, with an unhealthy appearance of the discharge, which becomes thinner, fœtid, or completely arrested.

It does not always follow that pus is found in the veins. However, the following case leaves no doubt in this respect; it is the most complete yet recorded on the subject.

It relates to an insurgent who received, on the 20th June, a ball in his left arm. The wound of entrance was in the outer and upper part of the insertion of the deltoid; the exit was a little above the level of the other orifice; the bone had been broken in pieces, several fragments were extracted, but still some adhered to the soft parts.

The patient went on well for the first few days. In spite of an application of gutta-percha splints they could not obtain union. *7th July*, the discharge becoming of an unhealthy character, they amputated the limb. The wound was closed with twisted suture. At some points union by first intention resulted, and both wound and discharge were healthy. *16th*, night sweats, no rigors, some weakness of the limbs. A little diptheritic sore throat, which was touched with alum; ordered a muriatic acid gargle. *17th*, Tongue dry, languid, a little shivering, sweating sickness, chest healthy.

18*th*, More rigors about eleven o'clock, otherwise the wound looks healthy; some white patches scattered through the mouth. 22*d*, Pretty well, no rigors, patches in mouth less. 23*d*, Increase of fever, cold, hot, shivering; ordered quinine.

24*th*, Pretty well; wound has a good look. A few days after this there was another accession of fever.

28*th*, Signs of effusion within the chest; delirium without suffering.

29*th*, Very ill; twitching of facial muscles. Died at two o'clock.

Autopsy.—In the stump, the cephalic and deltoid veins were full of pus as far as the axillary vein, which also contained pus, an abscess in the deltoid, others diffused along the biceps tendon, ascending into the joint, the cartilages eroded and removed at some parts of their circumference.

In the chest—on the left were adhesion, on the right a slight effusion, false membranes, an abscess on the lower surface, no abscess of the liver, attachments to its upper surface, nothing in the stomach, colon full of flatus.

In a great number of other cases, which we do not describe here, death ensued after two or three rigors, and neither the state of the wound or the organic symptoms accounted for the rapid termination.

Without knowing the results of anatomical investigation, one would, from the general appearance of the patient, suppose him to be labouring under the effects of an infectious poison.

If we have noticed, at our commencement, the old delusion which considered a gun-shot wound primarily a poisoned one, we cannot deny, on the other hand, that these wounds, when they are mortal, develop almost invariably a general condition, which is nothing more or less than blood poisoning.

But why, it may be asked, should gun-shot injuries have the power of doing this; and why, too, in such a much larger proportion than ordinary wounds?

The cause of the difference consists probably in, first of all, the sort of injury a ball causes. In all cases where there has been a post-mortem examination, have been found a centre more or less extensive, sometimes very large, of suppuration of all kinds, pieces of bone, the bullet or other foreign bodies lying in a bag of destroyed tissue. Wounds produced by other causes than fire-arms do not generally give origin to these foci of infection, because they are not generally accompanied by such a bruising of the soft parts which soften down and decompose before being eliminated, and by foreign bodies, especially bony splinters, which in a few days set up a profuse and ceaseless flow of pus, into which open a great number of absorbent vessels and veins not yet healed over.

The moral condition of the wounded man evidently influences the future of his wound. The insurgents in 1848 were found more difficult to heal than the defenders of public order.

We know, too, how the state of a soldier's health is affected, as he is the conqueror or the conquered.

"See," says Roux, "the sad spectacle which our wounded presented to us in 1814 and 1815; their moral depression by defeat, the privations of all kinds which they had suffered, delivered them easy victims to typhus and hospital gangrene."

Dr. Serrier, senior assistant-surgeon, speaks of a wounded soldier in 1834 at Metz, with merely an abrasion on the front of his leg. He was so affected by the description of an amputation some persons discussed before him, and which he thought applied to his own case, that cerebral symptoms appeared, and in two days the poor fellow died.

Another man was admitted in 1839 to the Hotel Dieu of Marseilles, for a gun-shot wound of his leg. Everything appeared to do well, when a visit from one of his daughters, who wept for two hours on his pillow, affected him so much that he died from mental excitement in a few days.

M. Baudens, in his " *Clinique des plaies d'armes à feu*," mentions a good example of the effect of moral influences, the case of an Arab, from whom he removed half the head of the humerus, and who, after the operation, continued to live with his own people, eating and drinking almost as if in health. Another Arab, whose fore-arm the same surgeon amputated, marched on foot for several days, refusing a place in the ambulance tent, preferring to pass the night in a café.

I myself remember a young " garde mobile " whose right arm was removed at the shoulder joint, who exchanged jokes with the surgeon while he removed a gangrenous flap, the size of a hand, hanging by a pedicle to the rest of the wound.

CHAPTER IV.

TREATMENT.

1. *Opening and Dilatation.*

SINCE it has been agreed to discard the idea that gun-shot wounds are poisonous, we must erase those rules which were of old employed with the view of destroying the effect of that imaginary poison. Who in our days would venture to introduce hot turpentine into a wound? Ambrose Paré, to whom we owe this reformation, speaks of the inflammation this irritating treatment caused, as the reason for his discarding it.

But there is one manœuvre which for long has played its part in military surgery, and the just value of which it is our duty to consider, viz., dilatation and enlargement of the orifice of a wound.

There was, in former days, a practice generally adopted, which consisted in enlarging, by incisions, the track of a wound, with the view of changing it into a simple one, and extracting any foreign body which might be present, and to prevent the extreme tension which, at a later period, they thought the tissues in the neighbourhood would be subjected to. Celsus recommended it to remove arrow heads, bits of stone, iron, etc. Now-a-days, we have generally given up this enlarging of wounds, except for some definite object, such as the extraction of splinters of bone too large to pass along the passage left by the ball, or to extract a foreign body, the presence of which would be hurtful; for instance, pieces of wood, which in sea fights often pierce the flesh. In these cases dilatation is clearly indicated; it is nothing more than an operation for the removal of a foreign body, not the dilatation for dilatation's sake alone, prescribed by ancient surgery.

Their object, giving exit to pus and sloughs, has no foundation in fact, for the opening made by the ball is generally sufficient to allow such debris or secretions to pass; were it not, the proper practice would be, not enlarging the original orifice, but making a suitable counter-opening.

It is another object the enlargement (debridement) is destined to fulfil. A gun-shot wound possessing all the characters of a *contused* wound, they say, dilate it with the knife, and it becomes a simple incised wound.

This explanation is merely theoretical, for at first the bruising of the tissues by the ball is wholly different to what we, under ordinary circumstances, call the contusions from ordinary causes. The velocity with which a ball penetrates produces solutions of continuity often resembling those made by a cutting instrument; so they heal by first intention, and there is seldom anything to prevent their doing so; it is not so much the special nature of the surfaces exposed as the necessity of allowing the debris to escape, and the impossibility of keeping the surfaces in apposition. So the wound made by the knife heals at first, but it comes to the same thing in the end; and it may be, on this account, that John Hunter argued against dilatation. The latter, then, cannot change the nature of the wound.

There is one object, however, which it is intended to attain—an object too important to be passed over silently, viz., freeing the tissues bound down and swollen by inflammation. This is a more specious reason. If the deeper incision could really obtain for the swollen tissues space in which they might distend, then it would become a surgical maxim.

But alas, facts do not support this theory, and freedom is but seldom obtained by this proceeding. Indeed, does not the true obstruction exist in the swelling of the tissues themselves?

It is not even to be found in the centre of the tissues swollen, and what new place could a deep incision discover for them to distend in?

Is it not, on the contrary, only the first wound which has made a space that no secondary one can appreciably increase? If it be the fibrous tissues which offer the resistance to free dilatation, is that obstacle met with in the midst of muscular tissue? It is only necessary to remember the anatomical structure of the muscles of the extremities to see that these fibrous tissues which could resist distension are more on the surface than the centre, and that it is the fascia which essentially binds them down and restrains them within definite limits. Besides, the inflammatory swelling does not confine itself to the neighbourhood of the wound, but often involves the whole thickness of the limb.

Is it not clear, then, that the enlargement in cases of muscular tension ought not to be at the deeper part of the wound but on the surface, and ought it not to consist then by cuts made into the fascia.

Such is, so far as we can see, the practice which should supersede the internal dilatation, viz.—

To make on the surface of an inflamed limb long incisions as deep as the fascia, in sufficient number to insure a freedom for the subjacent muscles.

This is a method so simple, safe, and rational, that it ought, in our opinion, to be practised whenever there seems reason to suspect tension from swelling. We have seen this practised many times in Paris, especially by M. Jobert, and with evident advantage. Muscles, after these extensive dilatations, formed true herniæ, and demonstrated, by their excessive enlargement, what would have occurred within the aponeurotic covering.

John Hunter, one of the first to set his face against these widenings of gun-shot wounds, founds his reasons principally from the following arguments :—

1. The number of cases which have got well without being dilated.

2. The new incision heals up before the natural expulsion of the foreign body can occur.

3. Inflammatory tension is not lessened by it, on

the contrary. The cases where it is admissible, under the reservations indicated, are as follow :—

1st. For the necessary removal of a foreign body.

2d. For securing a bleeding vessel.

3d. To remove a large splinter.

4th. When a vital organ is pressed upon, as the brain.

To replace some displaced organ, as pieces of the skull, ribs, bowels, lung, etc. etc.

In France M. Roux, with some lingering adherence to the principles of the old school, has shewn himself a partisan of the precautionary dilatation.

M. Begin advises that fascial wounds should be widely opened. M. Malgaigne is opposed to this practice. M. Velpeau denies its being of any use. M. Jobert calls it useless and dangerous. M. Baudens utterly casts it away. Dupuytren recommends incisions for the relief of inflammatory tension ; but in perusing this author's works, it is easy to see that he does not mean in general the dilatation of the wound itself, but rather incisions of a depth suitable to the wound through the fascia, in other words a thorough *freeing* of it (debridement veritable). He even notices the danger of dilating the deeper tissues. It was thus that a sick man at St. Cloud, whom he saw, was deprived of the use of his pectoralis major muscle by too deep a dilatation.

In the Crimea they but seldom enlarged wounds. M. Valette received at Constantinople wounded men from the Crimea, and kept them till cured ; out of one hundred wounds, involving fascia, all but one got well without either primary or secondary dilatation. Whence that author thinks that, as a primary step, it ought to be erased from military surgery, and as a secondary one it is rarely necessary.

M. Quesnoy considers, as one of the means for preventing over-distension of the tissues—what was found useful in the Crimea—viz., the advice given by all surgeons of the French army to their wounded, to keep the first dressings constantly moistened with cold water.

Unluckily, this important rule, observed so long as the wounded were in the ambulance, was often neglected during their transport from the Crimea to Constantinople. On their arrival they had some serious complications, which could perhaps have been prevented by an attention to detail, which, indeed, is often impracticable in war.

The dilatation of the wound, when it is deemed necessary, can be made with an ordinary bistoury, which will make a sufficiently but not too deep incision. In this case the instrument should be followed by the finger, on which it lies flat, then it is turned on its cutting edge so soon as it has reached the deepest tissues. Still the finger can guide and direct its movements. But should a deep incision be required, a straight probe-pointed bistoury should be introduced, with slight pressure from behind. The external dilatation consists in incisions of various lengths made with a sharp-pointed knife. Their depth is determined by the thickness of tissues they have to pierce.

One may make what number of incisions seems necessary, according to the intensity of the inflammation. I have seen cases where four or five longitudinal cuts in the thigh looked like the slashes in the sleeves of the middle ages.

Of course, it would be absurd to close these wounds, and would be neutralizing the good they are intended to accomplish.

This dilatation is, in general, only practised in wounds of the extremities. What, then, is to be done in wounds of the cavities? Here it would be more often hurtful than useful; and our remarks on the abuse of the probe in such situations, would apply to it either in the chest or the abdomen, fulfilling no useful end, but perhaps injuring some important blood-vessel, and, especially in the abdomen, would be liable to leave a tendency to hernia.

In intestinal wounds the safety, if there is any, depends on the adhesion which is established between the wound of the gut and the abdominal wall; dilatation can only interfere with the operations of nature. Of course, a case in which protrusion has already occurred is excepted.

2. REMOVAL OF FOREIGN BODIES.

Next to the dilatation, the most important step to consider is the extraction of foreign bodies. The *presence of the bullet* is almost always a complication of wounds with a single orifice.

It has been judged differently at different epochs in
the history of military surgery. At the time when they
considered gun-shot wounds poisonous, it was reasonable enough that they should be anxious to remove the
infectious body at any risk; hence, manœuvres of all
kinds were devised to effect this object, even at the
price of horrible sufferings.

At a later period, physiological researches into the
nature of gun-shot wounds lessened this dread of the
ball's presence, and they frequently left its expulsion
or enclosure to nature. Such, among others, was the
opinion of John Hunter, who has cast over many points
of military surgery the light of his genius and experience.

The destiny of a bullet left in a wound varies much
with circumstances. Sometimes it gradually approaches
the surface, and is cast out by a process of elimination,
as is described by Roux; sometimes it remains as a
veritable foreign body, hindering cure, and keeping up
inflammation and suppuration; at others, the wound
heals over it, and it may remain for any length of time
in the body. Sometimes it changes its place, and re-
appears far from its first resting-place.

It may happen, in some cases, that it occasions
alarming symptoms from pressure on a nerve, as I have
once remarked myself.

Larrey, in his Memoirs of Military Surgery, gives
in detail a description of a case very remarkable for
the gravity of the symptoms, caused by pressure of the
ball on a nerve in the neck, and their disappearance
after its removal.

M. Jobert tells of dissecting the arm of a man in whom a bullet had lain for twelve years, since the battle of Waterloo.

Marat relates, that a ball remained in the apex of a lung, without causing any accident. John Bell does not seem to consider extraction absolutely necessary, though he cites some cases where the foreign bodies hindered cure.

M. Manget and Diemerbrock tell of having known a woman who carried a ball in her thorax, which rolled about on the slightest movement. M. Malle cites the case of an officer wounded at Wagram, who attained a very advanced age, in whom they found a ball lying in the left lobe of the brain. M. Baudens speaks of a soldier wounded at Waterloo, who has kept for several years a ball in his frontal sinus without fracture of the inner wall.

M. Begin mentions examples proving the contrary to those just quoted, and which, he thinks, shew the necessity for extraction whenever it is feasible.

The opinion of this military surgeon agrees with the records of the Hotel des Invalides, for which I am indebted to M. Hutin, surgeon-in-chief to this great national institution, and inspector of military health. Here is his letter—

" We have still two or three soldiers of the first empire who carry about balls which struck them thirty-five years ago. They are very often incommoded by them ; but as they lie very deep, one does not dream of extracting them. During the twelve or thirteen years I have been at the head of the duties at this hospital, I have had to extract several which were superficially fixed and encysted, or which became too troublesome."

M. Valette says he was able to extract the ball in four cases with facility during the suppurative period. In two other cases, attempts at extraction caused acute inflammation; they abstained from making any, and the patients recovered, and carried about their bullets without inconvenience.

The conclusion which seems to result from these different opinions seems to be, that the bullet being neither more nor less dangerous than any other foreign body of the same shape, ought to be treated as such, that is to say, extracted when it is possible to do so without a dangerous operation, and left alone when such a proceeding seems too complicated. The extraction of other foreign bodies is subordinate to the possibility of performance, to their more or less angular form, to the accidents they appear able to cause.

A great number of instruments have been invented for the extraction of a bullet, but most of them have been given up by army surgeons.

They affect one of the three following shapes :—

1. Forceps.

2. Spoon.

3. Cork-screw or ramrod screw (tire-bowrie).

The first usually require a degree of lateral separation of the blades, very painful, nay, in some cases impossible, the second do not offer any purchase, and the three require a firm *point d'appui* which neither flesh nor bone afford them. Sometimes it is necessary, in order to remove the ball, to penetrate even to the bottom of the wound, which is extraction properly so called ; sometimes it is sufficient to make a simple incision, as if opening an abscess, the ball being lodged under the skin. In these cases it is generally found at a point *opposite* to where it entered, it has pierced the thickness of the limb, but in doing so exhausted its velocity.

Guthrie has thus extracted great numbers of balls by incisions an inch deep, and, contrary to the opinion of Hunter, no inconvenience resulted. Surgeons are not agreed as to what should be done with a *ball which has been driven into a bone.*

Probably this difference of opinion is due to the want of experience in the matter. It is in reality rare to find a ball penetrate into a bone without breaking it, and then the surgeon's attention turns to the fracture.

It is only in the spongy ends of bones that a ball can penetrate without diffusing its violence beyond where it strikes. The necrosis and caries which Guthrie attributes to the presence of a ball are perhaps owing to other organic disturbances which accompany the entrance of the projectile. The extraction in these cases is certainly advisable, but experience does not seem to confirm the extreme danger that author attributes to having a bullet in bony tissue. Trephining, recommended for these circumstances, has been generally employed in latter years.

Splinters of bone are, of all foreign bodies, those which present the most serious complication by their angular form and their hardness, so one cannot too urgently support the advice of most modern army surgeons, to remove all splinters as soon as possible, to separate even those which still adhere to the soft parts, and that before suppuration has had time to commence. I know by experience that it is not always easy to determine the presence of these foreign bodies in a wound so narrow as that caused by a ball. But we must not forget that the removal of splinters is much less painful in the first days than when the subject has passed through a tedious and weakening suppuration, and by weeks of pain and anguish his nervous sensibility is raised to the very highest. Considerations of humanity, apart from the surgical reasons, would suffice to impose on the surgeon the duties I allude to. But are not, for this immediate extraction, some peremptory reasons purely surgical, and especially that of not increasing without good reason, by the prolonged presence of splinters, a suppuration which pulls down the strength of the patient, so as to lesson and remove his chances from a secondary operation ? Who will not admit, for instance, with us, that the fracture of the femur in our first observation was much aggravated by tardy treatment.

To reduce as far as is possible by immediate extraction of the splinters a complicated to a simple wound is the great aphorism which military surgery places at the head of all rules relating to the treatment of gun-shot injuries to bone, in cases which admit of the preservation of a limb.

M. Baudens insists vehemently on the necessity of extracting at once as many splinters as possible. This maxim, founded on large experience of military surgery, did not find general favour in the hospitals after the days of June ; I do not remember having seen any case where it was rigorously carried into effect. M. Jobert expresses himself doubtfully, in these terms, " I never extract splinters at first (d'emblée), and I never touch openings which can allow them to escape no more than those which the projectile has made." Again, he says, " I confess that I do not hesitate to dissuade from the immediate extraction of foreign bodies as useless and even dangerous."

In opposition to this opinion, we again call to mind observation No. 1, of cure of fractures of the thigh. It is interesting to remark that in this case, treated in the Maison de Santé, which is one of the small number of cures recorded in 1848, they made an immediate removal of several large splinters which, being put together, made up nearly all the circumference of the bone to the length of three or four centimetres.

Professor Stromeyer, chief surgeon of the army of Schleswig in 1849, is satisfied of the importance of not submitting the fracture to too minute an examination, and of avoiding a long transport and a treatment which requires movement of the limb.

And he advises to readjust as much as possible any splinters not detached from their periosteum. It is astonishing, says he, with what ease fractures sometimes recover almost without exfoliation or removal of the splinters.

Since that date, the Crimean war has added its powerful testimony to that of the head surgeon of the Val de Grace, which may in other respects, I believe, be reconciled with that of the author we have just named. "The greatest care," says an English surgeon, building on his recollections of the Crimea, " should be taken at the outset to remove all splinters, all fragments adhering loosely to the tissues, which at a later date may prolong the suppuration."

French surgeons are equally agreed with their master of the Val de Grace.

For instance, M. Quesnoy says, "The removal of splinters is now-a-days an absolute necessity, demonstrated by experience. To remove all free bits of bone, to detach those small portions which still adhere by a jagged end, is the first thing to be done, and we have always followed, and intend to follow, this rule in the ambulances." Valette, of Constantinople, says he had but rarely to extract splinters, owing to the praiseworthy practice of his brethren in removing them on the field of battle. His remark is valuable, as it completes the observations of the field-surgeons, who could only follow up their observations for a few days.

The removal, as soon as possible, of all splinters has thus become a general rule among French military surgeons, and on which we have found no difference of opinion. It may be considered as now a part of surgery, and established upon large and undoubted experience.

It is probable that now it would be more generally followed in the Parisian hospitals.

3. FIRST DRESSING.

The first dressing of gun-shot wounds should be simple : a layer of charpie (plumasseau), and over that a compress dipped in cold water, retained by a broad bandage loosely applied ; and in the case of a fractured limb, to fix it by splints of gutta-percha, leather, or card-board. This enables the surgeon to allow of his patient being removed to the ambulance or hospital tent, till the time comes when all the wounded being removed, he has leisure to make a more careful examination into the nature of their injuries.

At the end of this work we shall give a description of the means of transport and treatment of fractured extremities.

Crimean experience has shewn the importance of closing the wound as rapidly and completely as possible, so as to avoid a sort of loss of nervous excitement which appears to result from the exposure of extensive raw surfaces.

It is very necessary that we should not apply a very tight bandage or irritating plaster over the wound. Our previous remarks on muscular engorgement requiring relief from incisions, shews sufficiently how important it is to allow for the inflammatory swelling of a wounded limb. When there is a fracture of a long bone, however, one may, in order to avoid shaking during the journey, apply the apparatus more firmly ; but so soon as the patient has got to his journey's end, you must loosen the bandages in such a way as to obtain immobility with freedom of expansion.

The dressing ordinarily employed by the English surgeons in the Crimea was more simple even than that of the French ; compresses wet with cold water, and light bandages, form the chief feature. Hot emollient poultices are out of the reach of the field-surgeon, they were universally employed in 1848 in the Parisian hospitals ; they obtained relief for the patient, but sometimes debility from the colliquative suppuration which they favour.

Sometimes it has been recommended to introduce tents into the wound to hinder cicatrization or false adhesions. This precaution is more theoretical than truly useful. It is very rare that cicatrization takes place while there are still sloughs or splinters at the bottom of the wound. Deep suppuration is in these cases so abundant that the opening it escapes by has no time to close. If, on the contrary, suppuration is so weak that the wound cicatrizes, one would not expose it artificially to an exhausting discharge with the object of being able at a future period to extract a splinter. The great danger of gun-shot wounds, setting aside the importance of the wounded organ, are the consequences of the suppuration ; that is to say, the exhaustion of the patient, necessity for painful operations to give exit to the pus, and so on even to purulent absorption ; if one could conduct the patient safely to the time when, suppuration diminishing, he can regain his strength, the operations to which he would have to submit for the removal of splinters, the union of broken bones, etc., would be rarely of so much importance as to compromise his safety. If, in consequence, one sees that at some period more or less distant one shall have to extract a foreign body or a splinter, it would be better to allow the wound of the soft parts to cicatrize, reserving to oneself the power of making a fresh incision when extraction is considered necessary.

But, let us not forget that these cases present them selves very rarely, and oftener it is precisely the presence of the foreign body which hinders all healing, and which in consequence most frequently claims all the attention of the surgeon.

Finally, the precaution of closing artificially an open gun-shot wound, appears to me more likely to do harm than good. As to the consecutive treatment, one cannot insist too much on the importance of favouring, by every possible means, the escape of pus ; and in the case of fracture, of avoiding every change of apparatus which is not absolutely necessary ; in short, to obtain for the limb as near as possible a condition of perfect immobility.

4. EMPLOYMENT OF WATER.

Cold water, in different ways of application, has been advised as the most antiphlogistic remedy.

Guthrie, in particular, recommends it in the form of wet compresses ; he allows, however, that in some cases there is an instinctive feeling against it, and in these it should be replaced by emollient poultices. It is really one of these questions which we dare not dogmatise or theorise upon. Like almost all questions of pure experience, it does not admit of an absolute general rule. In some cases I have seen cold water, and in particular, continuous irrigation, employed with evident advantage, through the suppurative period, to the actual healing of the wound. In others, again, the patients complained of rigors, of discomfort and local pains from this damp cold, and felt comforted so soon as it was replaced by an emollient poultice.

If I should venture to resume my observations on the employment of these two dressings, these are the rules which appear to me to correspond with my own experience. Cold water is applied with most advantage from the commencement, and before suppuration begins. It has then the advantage of resisting inflammatory reaction, and so preventing too profuse a suppuration, and of allaying pain.

The best form is undoubtedly continuous irrigation. This hydrotherapeutic treatment unites all the antiphlogistic advantages of water.

It shares with the bladder of water the continuous action of cold, but the latter assuredly has the additional advantage of evaporation, and exercising no pressure on the wound. A bladder half full can, however, be substituted for the latter sometimes, but the limb cannot always bear its weight ; it should always be suspended, so as only to act by its flattened surface.

Independently of its temperature, cold water can do harm by the damp which extends into the bed. This inconvenience may become serious, cause shivering, general nervous distress, and pulmonary or intestinal inflammations.

Cold water is particularly to be dreaded by soldiers subject to gout or rheumatic affections. Damp compresses especially present this danger, and ought only to be employed temporarily. They have, besides, the disagreeable necessity of having to be changed very often, and for this reason given over to the responsibility of a nurse or the patient himself.

Continuous irrigation, it is true, is sometimes as inconvenient, so far as dampness goes, and therefore is most adapted for parts distant from the trunk, as comminuted fractures of the forearm, hand, knee, leg, and foot. A partially filled bladder should replace it in wounds of the head, neck, shoulder, pelvis, and upper end of the thigh, where continuous irrigation is more difficult of application, and more liable to wet the rest of the body.

Extensive cooling with water was not in general use in the Crimea. They dreaded, especially in ambulances, the damp, the dangers which we have mentioned, and the impossibility of exercising the surveillance the treatment required. On the other hand, they generally advised the wounded to moisten the first dressing with cold water, up to the moment when suppuration begins, indeed, to encourage its coolness by continual evaporation.

Crimean experience shewed us, moreover, that suppuration is the limit beyond which cold water dressings should not be used.   Thus, M. Valette (we have
seen that he received the wounded into the hospital at
Pera from three to eight days after the event) congratu-
lates himself on having adhered to dry dressing.   He
had 10 dead out of 280 wounded.

*Continuous irrigation cannot be too much recommen-
ded in wounds accompanied by great breaking of bone,*
which lead one to dread intense and prolonged inflammation.

In these cases it possesses an undoubted preventive
usefulness in influencing, from its origin, the flame which
threatens to consume the limb, and to conduct it on its
course to cure, more slowly perhaps, but with less risk.

In support of the preceding statement, we shall
mention one case selected from several others.   It is
that of a young soldier received into the Maison de
Santé.   There were two wounds, one in front, the other
behind the external malleolus of the left side.   The
finger detected an extensive fracture of the bones of the
tarsus.

Up to the 14th July there were symptoms of intense local inflammation, but little general excitement.
Bleeding, leeches, poultices, were used.   From the
14th to the 23d repeated hemorrhage, considerable
swelling of the limb, greyish pouting wounds, extremely
painful.   From the 23d he was treated with continuous
irrigation ; under its influence the bleeding rapidly
ceased, swelling and pain diminished, the wound re-

gained its proper appearance, fever stopped. On the 5th August he could support himself on his leg, and they then stopped the irrigation, which had been continued for thirteen days. By the middle of October the wounds were healed completely, only an anchylosis of the ankle joint remained.

How long should one continue the employment of these cold applications? It is difficult to answer this question generally. I have seen wounds heal over while incessantly bathed. But, in general, the term can be placed at the time when suppuration is established.

The continuous employment of ice has been recommended by M. Baudens, who has made a special treatment of it, and assures us of having obtained surprising results. He distinguishes, with regard to tolerance of cold, *normal* from *pathological* caloric. He thinks that the process of inflammation develops caloric at the place affected, independently of that produced by respiration. That is what explains, according to him, how wounded limbs support, without being chilled internally, an amount of cold which would be intolerable to a healthy part.

Immediately the pathological caloric has been reduced to its normal quantity, the body has an unmistakeable sensation of it. That is the time to suspend the use of ice. The surgeon of the Val de Grace prefers ice applied for several days, or even weeks, to any other antiphlogistic ; according to him it is a depletion (saignée) of caloric. One may readily believe, from

F

what has been said, that poultices are but rarely used
at the Val de Grace, in gun-shot wounds. That would
be, says the author just quoted, to subject the integu-
ments to two macerations, that of the pus within and
the poultice without.

The manner of applying ice consists in laying the
injured limb on a cushion, guarded by a waterproof
cloth. The wound is covered with a light layer of
charpie, on which they place the pieces of ice. This
method is peculiar to the eminent author just quoted.

The surgeon-in-chief of the army of Schleswig, in
1849, declares himself in favour of cold water, and
deplores not having ice at his disposal. M. Roux
has not expressed himself favourably in regard to
refrigerants. M. Velpeau is strongly opposed to cold
water, for the reason that it arrests the inflammatory
process necessary to proper cicatrization.

Finally, we ought not to pass over an employment
of water directly opposed to M. Baudens', viz., irrigations of water slightly tepid, recommended by M.
Amussat as replacing poultices with advantage, and
even the antiphlogistic effect of cold water. We have
no experience of that method, and we do not wish to
express a decided opinion on hearsay.

The state of the patient, whether it is more or
less sanguineous, his age, as well as the season, ought
to influence the employment of water. A vigorous
young subject, having lost little blood, wounded during
the heat of July, will require undoubtedly, especially
during the first eight days, a cold temperature, or even

ice. A man already weakened by age, hemorrhage, or during a cold season, will have no need of such energetic local antiphlogistic treatment; then water at its ordinary temperature, or tepid, will be most suitable. We must bear in mind, that water acts not only by its cold but by its evaporation also, which is naturally more rapid when the water is heated. We may judge of the importance of evaporation, by recalling to mind the heat which is developed in compresses, covered by an impermeable envelope. But it is very necessary to keep one's theories in subjection, and to proceed on an impartial appreciation of the particular case under our especial observation. Every one is agreed as to the utility of water dressing. The differences of opinion merely relate to its temperature, the duration of and time for its employment.

These differences refer undoubtedly, also, in addition to preconceived ideas, to the different circumstances in which the surgeons have been placed, to the extreme variety of the wounds confided to their care, as well as to the variable predispositions in their patients.

Water is contra-indicated whenever the patient finds a general discomfort from it, shiverings, and especially when it favours inflammatory strangulation. The general opinion is, that this mechanical or vital constriction favours the development of tetanus, the former being combated with incisions, the latter by soothing applications. These last ought to be preferred to the cold, when the wound is seated in an inextensible tissue yielding with difficulty to the inflammatory swelling; they have also an anæsthetic effect, much liked by the patients when pain is acute and incessant, rendering sleep impossible.

Cold, moreover, is directly contra-indicated in all cases when the congestion of the limb makes one dread gangrene, or when the latter is favoured by the ligature of an important vessel.   Cold applications, warm soothing emollients, thus supply each other's place, or succeed to each other in turn, and it is for the tact of the surgeon to choose which to employ in each particular case.

These two agents are the first remedies to be employed in gun-shot wounds.

There is one besides which it is necessary to submit to an impartial examination; I allude to bleeding.   *In all things it is necessary to consider the end,* is a proverb which is very apropos to this question.   What is the special character of the wounds of which we speak ? Loss of substance and contusion of neighbouring parts, and as to bones, fractures with numerous splinters. All these circumstances are favourable to the production of abundant suppuration.   Gun-shot wounds which heal by first intention are rare, and always slight.

In general, cure, especially in cases of fracture, only takes place after abundant suppuration, and consequently after great general debility.   How important must it be, then, to preserve for that time all the vital energies, and not to weaken them during the inflammatory period by repeated and copious bleedings ! The cure of these wounds not being common without the elimination of debris or foreign bodies, the inflammation is in the majority of cases only a symptom of this process of expulsion (travail d'expulsion), and cannot, in consequence, be treated so efficiently by general antiphlogistic remedies as an idiopathic inflammation. The true antiphlogistan of fire-arms is the removal, as completely as possible, of every kind of foreign bodies. These reasons for our dread of bleeding do not prevent us from recognising its value in cases where, in consequence of some individual peculiarity, there is established an inflammation disproportioned to the gravity of the lesion.

After the revolution of June 1848, bleeding was only practised where there was a positive indication for it, but not as a precautionary measure.

They even sought to prevent exhaustion of strength by prescribing a nourishing diet from the first.

M. Velpeau, in the discussions of the Academy of Medicine in 1848 was equally strong against preventive bleeding, and only reserves it for the time when the exigencies of the case required it.

Besides, it is certain that sometimes gun-shot wounds heal without any evident inflammation, and then precautionary bleeding has no rational object. M. Huguier is of the same opinion.

M. Jobert, on the other hand, advises abundant losses of blood by general bleeding in the inflammatory period.

Professor Stromeyer advises bleeding, leeches, and cooling diet. He, among others, has seen that treatment diminish suppuration. Dupuytren recommends several bleedings ; each to be copious and repeated. Larrey frequently employed it.   Baudens endeavours to substitute ice.

In the Crimea, bleeding, either as a preventive or a cure, was very rarely employed.

Every one has felt the importance of avoiding, in the treatment of a case, all causes of debility.   There, wounds of the chest were excepted.   For them, Macleod, among others, recommends bleeding, after his Crimean experiences ; perhaps, too, from his remembrance of John Hunter's precepts.   Is it not probable that the greater care to avoid losses of blood in the war in the East than at Paris depends, to a great extent, on the difference of the hygienic circumstances in which the wounded were placed ?   From what point can we fully appreciate the condition of a wounded man after a great campaign ?

Is it not to be admitted that his probable state presents a general deterioration and debility of his system and an alteration of his blood to a greater or less extent ?   Typhus, scurvy, fever, cholera, hospital gangrene—which caused such havoc in the Crimea, since, of 300,000 men, there were 200,000 sick, 150,000 of whom were unwounded—do not these affections depend on a decomposition of the blood, a sort of dyscrasia peculiar to camps ?   That dyscrasia exists, indeed, even when there is no outward and visible sign of it.   Thus, as we have said, among many wounded in the Crimea, after death were found ulcerations of the intestines analogous to those of typhus, as if their bodies had been insensibly *typhoidised;* along with that, the wounds had very often a pallid tint, and healing was accomplished incompletely and without vigour.

What advantage *can* one derive from bleeding in these cases? What influence *can* it exercise on a fever depending more on debility than on a true vigorous inflammation?

After the days of June it was entirely different; the ball in general struck a body in robust health, which had undergone no fatigue except that of political excitement. So bleeding was somewhat more rational, and under these circumstances it would seem that one may have to follow another form of treatment to that pursued in the Crimea.

5. Diet and Hygienic Measures.

If the safety of an entire army depends above all things on the hygienic principles according to which it is directed, and if, in this sense, one may say with reason, " Feeding is the whole of a man," I venture in my turn, and with conviction, to add, *" Feeding is the half of a wounded soldier."* There seems a general agreement now-a-days to lay aside absolute rules of diet in the treatment of cases, or even a severe regimen. To this question will apply the remarks we made on antiphlogistics.

Let us not forget that in complicated wounds it is necessary, for the first eight days, to prepare the patient

for the enormous loss which will occur during long
weeks of suppuration.

One must see, as we have seen on a great scale, to
what a degree the strongest bodies can be attenuated
and reduced to skeletons by the suppuration of complicated wounds, to appreciate the importance of husband-
ing strength from the very outset.

That is the point which in general guided French
surgeons after the battles of June 1848.   Whether
that opinion was somewhat modified, or that one was
disposed to profit by the immense resources offered by
the town of Paris, the diet was generally nutritive from
the first.   Professor Malgaigne declared most vehemently
on this point ; he gave even wine to his wounded immediately after the inflammatory fever of the first few
days.   Velpeau agrees with him.   He does not see, any
more than the author just quoted, why a comminuted
fracture of a limb should deprive the man struck of an
amount of food proportionate to his appetite.   We shall
not increase the danger to the general health by allowing of the development of the local excitement due to
the process of elimination and cicatrization, and let us
not imagine that a lowering regimen better favours the
progress of this perilous but beneficent struggle of the
body with the local injury.

All the experiences of the Crimean surgeons go to
prove the importance of a good nourishing regimen for
the preservation of an army.

The English fed their wounded more abundantly
than the French, except when the wound involved any

of the great cavities, and they did not preoccupy themselves so much with the dangers of inflammation.

They boasted, moreover, of the use of cod liver oil in cases of great suppuration.

No army is fed like the English army; none, moreover, is better clothed. Thus, in preserving her soldiers, England makes up for difficulty of recruiting.

Here is the ration of an English soldier on a campaign—

Biscuit . . .	lb. i
Fresh meat . . .	lb. i¼
Salt do. . . .	lb. i
Rice . . .	℥ij
Sugar . . .	℥i¼
Coffee . . .	℥i¾
Rum . . .	a gill.
Charcoal . . .	lb. i½
Light . . .	℥ij
Tea . . .	℥¼

The bread is less than in the French army, but the ration of meat is almost double.

Here is in fact the ration of the French soldier—

Meat . .	300 Grammes.
Wine . .	46 Centilitres at two distributions.
Coffee . .	20 Grammes.
Sugar . .	25 do

According to M. Malgaigne, the Russian diet roll was for the half portion.

Bread . .	½ Kilogramme.

Meat .	.	.	200 Grammes.
Vegetables	.	.	120 do.
Wine .	.	.	½ Litre.
Brandy	.	.	1 Decilitre.

According to Dr. Armand, attached to the ambulance of the Imperial guard, the French army could not have fared less well than the English, so far as the diet of the wounded goes. The physicians had, in this respect, a great latitude.

Boiled meats, soups, wine, coffee, chocolate, milk, and fruits were, according to the medical prescription, placed at the disposal of the patients.

6. HEMORRHAGE.

Bleeding, which accompanies the wounds of which we speak, may occur under three different circumstances—

1. *Immediate* hemorrhage from the rupture of an important vessel, is the first accident we are called upon to contend with.

2. *Consecutive*, resulting from the softening action of the secretions of the wound on the clot which closed the opening of the artery, and on the cicatrix still tender and spongy.

3. Bleeding, not from any particular vessel, but the whole surface of a wound, as from a sponge ; it is clear, and of a serous character.

The primary, or immediate hemorrhage, generally

occurs on the spot, and before the arrival of the surgeon. They often neglect to offer at once the necessary succour, and when the latter arrive, the weakness resulting is already sufficient to compromise the success of the after treatment. It is important, consequently, that all those likely to be called to assist the wounded should be practised in the primary means of arresting it. The most important is placing the fingers in the wound, and keeping them there during the transport of the wounded from the battle field to the hospital. Amussat, who has particularly recommended that precaution, cites the case of a comminuted fracture of the right thigh with complete rupture of the popliteal artery and vein ; gangrene of the foot, followed by amputation, and death in seventeen days.

He attributes this fatal issue essentially to the profuse hemorrhage which took place at the moment of injury, and which could have been prevented by the means indicated.

The Russian army in the Crimea employed pensioners to afford the first help to the wounded till a surgeon could arrive. This plan was found very useful. I shall mention another instance to shew how important it is to bring help at once to one who is bleeding. Dupuytren mentions a case of a young man who, in 1830, received a ball in his thigh which opened his femoral artery. He lost so much blood while they were carrying him to l'Hotel Dieu, that he arrived there dying.

The femoral artery was immediately tied, but he

sank into a syncope, from which he had not strength to rally, and died.

By means of the finger introduced into the wound, one can appreciate, with intelligence, the degree of compression necessary to arrest the bleeding, and press directly on the trunk, or better still, on the orifice in the artery, which can be ascertained by the jet of blood.

One ought afterwards to favour the formation of the clot by direct compression of the artery with a tourniquet, and to make, according to Amussat, continuous irrigations of water, not too cold, over the wounded limb.

It need scarcely be said that during this transport of the patient all shaking should be avoided, and that it is necessary to keep constantly on watch during the period when a return of the bleeding is to be feared.

If the hemorrhage has been so severe as only to yield to a strong pressure from a finger or hard compress, it is probable that an artery of considerable size has been injured; and in this case it is necessary at once to apply a ligature.

No plugging then, with or without a styptic, no tourniquet furnishes then a sufficient safeguard, and neglecting the only perfectly sure means of arresting hemorrhage would be to expose the patient to the greatest peril. In these cases one need not fear inducing vascular debility of the limb below the ligature, because the artery has been already ruptured by the ball.

Secondary bleedings are generally even more serious

than primary ones ; they occur when the patient has been already enfeebled by suppuration, and in so sudden and unexpected a manner, that often assistance does not arrive till he has lost a great deal of blood. Moreover, secondary hemorrhage need not be very profuse to cause death. It is reported to have occurred after a loss of not more than twelve ounces. In these cases the first assistance consists in compression of the main artery *above* the wound by tourniquet or finger.

This means is more efficient in secondary hemorrhage, because the circulation has then already lost its first vigour. If it be not sufficient, or if the region does not allow of its application, it is necessary to have recourse to the ligature. Sometimes the source of the hemorrhage is hidden in a cavity, and then the ligature is impossible ; if an external bleeding is arrested by compression, it is thereby converted into an internal one. The most rational hæmostatic means in these almost desperate cases, consists in introducing into the wound a piece of lint shaped like the finger of a glove, then to fill its pouch with charpie or cotton wool, so that when traction is made, it may exercise a pressure from within outwards.

Hemorrhage may produce its effects (éclater) very tardily, *e.g.*, a young man with a transverse gun-shot wound of the neck died in a few hours of bleeding, which occurred during a paroxysm of mirth when the wound was almost healed. The so-called styptic powders and waters, even the perchloride of iron, are inefficacious in the two first kinds of hemorrhage. The oozing (forme

parenchymateuse), on the contrary, can be arrested by
these applications, such as powdered alum, charcoal, or
quinine ; lemon juice, and especially perchloride of iron.

There do not exist, so far as I know, any statistics stating the relative frequency of these different
kinds of bleedings ; the second variety is the one I
have most commonly seen, and that is to be expected.
The ball carrying along with it a mass of flesh of its
own size, crushes the surrounding parts.   That circumstance, as well as the presence of debris not yet softened down, serves to stop up for a few days the openings in the blood vessels.   At a later date, on the contrary, the softening of inflammation from pus, sloughing, gangrenous particles, favour the re-opening of the
vessels, and bring about very formidable secondary
hemorrhages.

The following observations contain the classifica-
tion of hemorrhages, as suggested by our own observations :—

1. Ball entered near the umbilicus, escaped on the left
    thigh.   Profuse primary hemorrhage through the
    latter aperture ; death same day.

2. Ball entered under right clavicle ; exit at sub-spinous
    fossa.   Profuse hemorrhage 28th day ; died on
    that day.

3. Ball entered on left forearm, exit above the elbow.
    Fracture of ulna ; first occurrence of bleeding,
    27th day—compression ; second hemorrhage, 28th
    day—compression ; third hemorrhage, 30th day
    —compression.   Cured, and limb saved.

4. Penetrating wound of the chest ; fracture of 5th rib ; rupture of intercostal artery. Primary hemorrhage difficult to stop ; death on 13th day.
5. Wound of left thigh, without fracture. Profuse bleeding on 7th day ; ligature of femoral artery. Cured.
6. Ball traversed left thigh without fracture. Profuse hemorrhage on 29th day ; compression found insufficient ; ligature of the femoral at the fold of the groin ; two slight bleedings on the 36th day, ligature of external iliac ; fourth slight bleeding on the 42d day ; death on the 43d.
7. Wound of the right leg, with a comminuted fracture of the tibia. A flow of blackish blood about the 23d day ; probably from the anterior tibial artery ; in a critical condition, but terminated happily nevertheless.
8. Ball entered the upper third of the sternum, breaking that bone and one rib. The finger could measure the extent of the destruction. By the open wound in the sternum an abundant arterial hemorrhage escaped, the source of which it was impossible to determine ; as impossible was it to apply a ligature ; graduated compresses of Amadon, supported by a bandage round the chest, but the hemorrhage is still very likely going on internally, and death seemed imminent.

As may be seen, of these eight cases only two are of primary hemorrhage, the others occurred generally between the 20th and 40th days.

Dupuytren, John Bell, and other surgeons, agree with our observations in finding primary hemorrhage more rare than secondary, and in considering the latter the most serious. The ligature is less easily and surely applied to an artery when the wound has been subjected to inflammatory action. The vessel is retracted and more difficult to reach, the thread cuts more easily, and sometimes falls off before obliteration has been completed. In these cases the only means of safety is applying the ligature at some distance above the wound. Dupuytren performed this operation successfully on the femoral, carotid, and temporal arteries.

M. Begin has satisfied himself, in one remarkable case, that serious bleeding may occur between the wound and seat of the ligature, and that it is unnecessary to place it any farther from the wound. He argues from a disarticulation of the right arm, with a precautionary ligature of the axillary artery—hardly was the wound reunited than a gush of arterial blood burst forth, and was only stopped by a second ligature applied lower down.

The coldness of a limb after the ligature of an artery may sometimes be only a passing symptom, nevertheless ligature of the femoral, of which we possess three observations, is a very serious operation, and one which has been generally followed by mortification of the limb and death. The following observations require to be given entire, on account of the importance of the particulars, especially as relates to the ligature.

1. *June* 5, 1848.—A ball having traversed the left

thigh, without breaking the bone, penetrated into the middle third of the inguinal triangle, to find an exit at the lower part of the hip. The patient suffered much during the first few days. A profuse hemorrhage occurred on the 4th July. Compression and cold not sufficing, the femoral was tied at the fold of the groin. There was nothing particular about the operation.

For an hour the temperature of the limb was lowered, then it returned to an equality with its fellow. The patient felt a sensation of numbness. He did not appear to feel his leg; pulse, 120; anæmia, headache, discoloured lips.

July 6th. Bleeding from the hip stopped by compression. It occurred again in the morning, then again on the evening of the 11th; the external iliac was then tied. The 12th to 16th doing well, though pale and flabby. 17th, died.

Autopsy.—Peritonitis characterized by false membranes, stretching from the iliac crest, matting the intestines into loops. At the seat of the ligature, on the femoral, was an ulceration of the artery about an inch long, a clot adhering to its walls; at this point the ligature remained; above, a reddish clot, nearly an inch long, ending in a fine point, passed upwards into the internal and common iliacs adherent to the artery. As for the bullet, it had passed before the neck of the femur, avoiding the joint, turned round the small trochanter, breaking it, and shivering the tuberosity of the ischium. In this neighbourhood was a cavity, in the midst of which was the sciatic nerve, destroyed by suppuration, and steeped in blackish-coloured debris. In the sheath of the sartorius was a collection of pus. The knee-joint even of the same side was full of pus, but without alteration of the synovial membrane and cartilages.

2. 28*th June*. A gun-shot wound of the outer and middle part of the left thigh ; the ball escaped at the same level, on the inner side, without breaking the bone, and slightly grazed the other thigh. From the first there was insensibility, and a difficulty in moving the leg. The skin was insensible throughout, except on its internal surface. These symptoms disappeared partially when the wound began to suppurate, and since then they were concentrated on the foot, of which there was complete insensibility upon the 5th July. The patient could not move his toes save by the tibio-tarsal joint, and even that not without suffering.

The 5*th*, Bleeding, tolerably profuse from the internal wound, rendered necessary a ligature of the femoral at the fold of the groin. The limb did not grow cold, but the patient had no feeling in it, it was pallid and damp. *July* 8*th*, Doing well ; 17*th*, the ligature came away in the dressings, no bleeding ; 29*th*, well ; sharp pains in the foot, and along the sciatic nerve ; to be blistered along its course. *August* 1*st*, The patient is better, but still has pain in his foot; 14*th*, doing well; wounds closed, foot still painful, and difficult to move; leg a little flexed ; he cannot support himself on it, but walks on crutches. *September* 7*th*, He can lean his

heel on the ground, and for a moment rest his weight
on the foot if the leg be flexed; except at the ankle
the foot is insensible, the leg has sensation, but
its volume is a centimetre and a half less than the
other.

3. Fracture of the left thigh by a bullet which remained in the limb. The subsequent bleeding necessitated ligature of the femoral. In consequence, general
gangrene, blue in the lower part, passing into a red as
it ascended to the middle of the thigh; several large gangrenous bullæ. Yesterday even, they determined to
leave the poor fellow to his fate, not doubting that
the gangrene would ascend to the seat of the ligature.
Contrary to our expectation, that did not take place,
and on July 14th they still contemplated an amputation of the thigh.

From the superior flap there escaped with crepitus
numerous bubbles of gas from all sides, as from an incised lung, a putrid smell from the muscles, infiltrated
with black decomposed blood. From a muscular
arterial twig flowed a small quantity of black foetid
blood, one suture only required; death next day.

7. TETANUS.

It is curious that out of about a thousand wounded
which we had the opportunity of observing in the Parisian
Hospitals, we did not meet *one* case of Tetanus.

In the Maison de Santé, of eighty-four wounded
there was not one case in M. Roux's practice; in that of

M. Giraldès, at the hospital of the faculty, only one fatal case. That excessive rarity of tetanus should be attributed in the first place to the readiness and excellence of the attendance afforded in the French capital. Undoubtedly attention delayed, or given in an irregular and imperfect way, with equally deficient hygienic arrangement, and sometimes the want of necessary food till the exhaustion of strength occurs which often follows great battles, may be so many predisposing causes of tetanus, which fortunately were wanting among the wounded of the revolution in 1848. There is another cause which may have contributed to it, namely, an elevated temperature. Then, again, we know that tetanus is due more especially to the sudden changes of temperature met with in autumn. A list made by Professor Laurie of Glasgow tells—

> In summer - 48 cases.
> In the temperate season 51 „
> In winter - - 66 „

The table relating only to the town of Glasgow alone gives—

> In summer, 14.
> In autumn, 21.
> In winter, 15.

Seeing the scarcity of cases of tetanus in 1848, I am obliged to have recourse in this serious complication to my observations made out on the fields of battle.

I have seen a young man of seventeen, whose scalp had been almost torn off from before backwards by the bite of a dog, from tetanus, lasting from 14th to 22d November. Another young man of fifteen succumbed to the same disease, owing to a fall on his knee (he died in the month of *May*). A third was also a victim, though the wound merely consisted of a prick in the hip (November). A fourth had a broken thigh and a contused skull, followed by trismus, which did not become general, and he recovered. A fifth case is that of a young girl who had two teeth drawn. The lockjaw was accompanied by a painful condition of the cervical region, aggravated by pressure.

The tetanic condition was not general ; she gradually got well by the application of leeches to the back of the neck, frictions with laudanum, calomel, and repeated tepid baths.

This case differed from the others in the less degree of contraction of the masseter and cervical muscles, and especially by the spasms not being so general.

It is probable that the latter was the chief cause of her recovery.

Every one knows that tetanus essentially consists in continuous muscular contractions, but intermitting from one set of muscles to all the muscles of the body. It commences commonly by a slight stiffness of the jaws and difficult deglutition ; sometimes it is the injured limb, or the feet, or the neck, which are first seized by the spasms ; the muscles are then hard to the feel, their tendons prominent, pressure causes pain and stimulates to stronger contraction. The muscular tension, which is more or less permanent, becomes more violent momentarily and by sudden paroxysms. The spasm invades the muscular systems one by one, and thus places the whole system in a condition of convulsive tension more and more agonizing and painful. The condition then presents some similarity to hydrophobia ; swallowing is almost impossible, the masseters keep the jaws closed, and the forced pressure of one row of teeth on the other is very painful ; so also the contraction of the muscles of the neck drags the head forcibly back. There is sometimes an intermission from exhaustion, or a moment of sleep, but that is but a fugitive phase of the disease, all the alarming conditions reassume their first intensity, but when the remission is complete, then the muscles of the limbs and neck, etc., regain their usual softness, then the lower jaw falls easily from the upper; that is no longer a fortunate crisis in the disease, it is death approaching to deliver for ever the sufferer from his agonies.

The occasional causes of tetanus are, according to the opinion of both ancient and modern surgeons, sudden changes of temperature and cold, especially when these causes act upon an abundant suppuration and a profuse perspiration.

Larrey, to whom we owe the most important memoir of military surgery on tetanus, attributes almost all these cases to the action of cold and damp air. At one time it is the Nile, which bathes for three months the walls of the Cairo hospital, or sojourn in tents on damp ground, at another the rainy season, the evening breeze, or the chill of night. Tetanus, as a complication of gun-shot wounds, is almost always mortal. The cure of a general case of acute tetanus is an exception. A sergeant attacked by it in December 1800, apparently benefited by nine applications of the actual cautery, but he died on the seventh day. After the battle of the Pyramids there were five cases, and five deaths between the third and fifth days ; after the revolution at Cairo, seven cases, seven deaths within a few days ; after the battle of El-Alrich eight cases, eight deaths took place between the fifth and seventh days. At the taking of Jaffa were several cases, all of which died in two or three days. General Daumartin died on the sixth day. After the battle of Aboukir, ten cases, and as many deaths at the same period of the disease. Dupuytren does not mention a single recovery, and relates two fatal cases to prove the inefficacy of amputation. It really seems to one reading his memoir that he had never seen a cure.

In spite of these startling facts, which would make one think that death was inevitable from an attack of tetanus, there are recorded some undoubted cases of cure. Of six cases that we have observed, two were cured ; but we must remember they never reached the very severe stages. Larrey has cured several cases ; a Mameluke, æt. 27, had a complicated wound of the right hand ; neglected for the first few days, he was seized with tetanus, which was only properly treated on the third day of its appearance. Three days later he was almost well, when some imprudence brought on a relapse, from which they nevertheless rescued him.

General Lannes received a bullet wound in the leg at the battle of Aboukir; he was seized with tetanus, consisting of spasms and drawing up of all his limbs, clenched jaws, difficult deglutition, intense fever. His recovery, and an abundant suppuration from the wound, were coincident. M. Esteve, a friend of Larrey's, had the same symptoms in a still more intense degree, by the effect of a fish-bone sticking in the back of his mouth; he made a decided and rapid recovery after a profuse perspiration.

One amputation case was seized with tetanic symptoms, coincident with the suppression of suppuration and perspiration. The former was brought on again by the application of a large blister to the whole circumference of the stump, and then all the symptoms disappeared.

Another amputation case was cured of tetanus by powder of cantharides sprinkled on the wound, which established an abundant discharge.

It would seem from numerous observations that the male sex is more subject to tetanus than the female; of 41 cases, 13 were caused by burns; of these, 11 were women. Ages vary in this series between 4 and 63.

Ages of both Sexes.

	Years.	Cases.
Between	60—70 . . .	1
„	50—60 . . .	1
„	40—50 . . .	2

Years.			Cases.
Between 30—40	.	. .	7
„ 20—30	.	. .	6
„ 10—20	.	. .	13
„ 10	.	. .	5

The more slowly tetanus comes on, the greater are the chances of recovery. The shortest interval was three days ; the longest one month.

A rapid and intense diffusion of it over the body (generalisation) is also a bad symptom. If the tetanus continues over the first week, the chances are more favourable, all the deaths occurred before the seventh day.

Tetanus may run an *acute* or a chronic course.

It is oftenest traumatic. Some believe, even, that in the cases where there is no visible traumatic cause for the attack, that there was one, nevertheless, though it escaped notice, as one sees in hydrophobia.

However, it appears evident to us that it may be idiopathic. ˙I have one observation which I investigated with all possible exactitude, and which left no doubt as to the non-traumatic origin of the attack.

A man æt. 40 caught cold travelling by railroad. No wound. Immediately after, *6th January*, trismus, stiffness of the cheeks, pains in the neck, fever, violent perspiration. *7th*, Acute nuchal pains, peculiar expression, abundant sweating, pulse 120, the jaws would with trouble separate enough to allow protrusion of the tongue, head bent back, frequent opisthotonos, painful contractions of the epigastrium . . . . . bleeding, leeches, blisters, calomel and opium; in four days took 31 grains of opium. 13*th,* Very weak, facies hippocratica; died, 1 P.M., having taken 59½ grains of opium in six days.

P. M.—Spine, ½ oz. of rose-coloured fluid within the dura mater, latter of a uniform rose tint in its whole length; that of the brain was of a pearly white, contrasting with the other; arachnoid slightly injected, especially in the cervical region; cord perfectly white, without internal vascularity, no red dotting, medulla oblongata perfectly normal, cauda equina a clear white. Heart evidently softened, intestines lined with an inodorous brown mass, evidently the remains of the opium. Spleen reduced to a *bouillie.* A chemical analysis of the brown mass detected neither morphine nor meconic acid.

It is very difficult to get a clear notion of the special nature of tetanus. It is sufficient to look over the results of post-mortem examinations to be convinced that tetanus must be classed with these maladies—hydrophobia, eclampsia, narcotic poisonings, etc.—which baffle the sagacity of the best pathologists by the unsatisfactory results of their investigations, compared with the acuteness of the symptoms before death, and the extreme mortality. In general, indeed, autopsy only discovers a slight congestion of the cerebral and spinal membranes, sometimes a limited softening, and once only the signs of an apoplectic attack, which had complicated the tetanus.

How, in the face of these facts, is it possible to avoid the supposition of the existence of a *general* unknown cause, of some poison analogous to purulent infection, which, starting from some given point, extends gradually over the whole body, and especially to the nervous centres? It is clear that tetanus is not due to a diffuse inflammation of the spinal coverings, but that the inflammation, if it exist, is dependent on an unknown cause, which the ancients would perhaps have baptised by the name of *acre tetanicum*.

Tetanus, as every one knows, is generally accompanied by great restlessness ; the patient, even before it comes on, is agitated, anxious about the consequences of his wound, has a troubled expression, and suffers habitually from sleeplessness.

But if tetanus exalts nervous sensibility, the nervous disturbance in its turn reacts in the most unfavourable manner on the progress of the case, and that is perhaps one of the causes which, after the battles of which Larrey speaks, made tetanus a more frequent symptom than after the bloody struggles of the days of June. What must be the effect of the nervous exhaustion after a prolonged engagement, which has itself been preceded by forced marches and all sorts of hardships !

Although the wounded in Paris were excited by political emotions, they were not in such a condition. The bloody struggle which took place in the streets was not preceded in general by bodily exercise and privations, likely to exhaust the nervous system ; then, again, are we not aware how much the morbid predisposition to typhoid fever is increased by nervous depression? I have seen a mother, thirty years old, sink in four days from a frightful typhus fever, the result of incessant care of her sucking infant for two months.

Five days before her death, that constant mother was still on her feet, though tremulous, at the bedside of her child, pouring out for him the vestiges of vigorous health, but exhausted to the uttermost.

The state of nervous exaltation which accompanies tetanus is made manifest by the patients' tolerance, without narcotism, of enormous doses of opium.

Dupuytren confirms that experience. He has prescribed opium even to the amount of an ounce in three or four days without staying or changing the progress of the disease. Besides the catarrhal and nervous predispositions to tetanus, there is a third cause often indicated — I mean the character of the wound. It is generally thought that *contused and lacerated* wounds predispose to tetanus.

Our experience confirms that opinion—

1. Scalp torn by a dog.
2. Wound and fracture of the knee.
3. Extraction of two teeth.
4. Fracture of femur and contusion of right side of the cranium from a fall.

The tables drawn up in England prove the truth of this assertion by still further evidence :—

9 contused wounds of the hand and arm with gangrene,

1 comminuted fracture of the leg,

1 old ulcer of the leg,

1 suppurating wound of the eye,

6 lacerations of the back and leg,

1 laceration of the foot,

4 punctures of the foot,

2 wounds of the face, with fractures and foreign bodies,

1 laceration of scalp,

10 burns,

2 amputations at the ankle joint, one of which was primary—are the only cases from simple wounds.

Dupuytren, Larrey, etc., generally consider as a predisposing cause the presence of a foreign body in the wound, or the irritation produced by splinters of bone, also including a nerve in a ligature or involving it in inflammatory swelling. The English lists often allude to nerves found softened, bruised, or injected, in the wound.

The action of the various causes seems to us readily explained by the physiology of the cord, especially by its reflex action. A limb is irritated by something pricking or bruising it; in consequence of that local irritation arise motor actions, and the muscles contract more or less generally. Are we not assisted in coming to a proper idea of this by experiments on reflex action? By irritating the skin of an animal after decapitation one determines a current centripetal to the spinal cord, and thence a centrifugal action to one or several muscles. May not that same law explain also how the removal

of a foreign body from the wound, the action of emollients on the latter, and the re-establishment of arrested
suppuration (Larrey), contribute to a cure by diminishing the centripetal action from the focus of irritation?
Is it not indeed to the same primary law that one can
attribute the disastrous action of the nervous exaltation
of the subject, and the importance of removing all cause
of excitement in keeping him in darkness, and avoiding
all sudden emotions?

We may recall to mind the condition of an animal
whose reflex action has been exalted by nux vomica.
It is often sufficient to tap gently the table on which it
lies, to touch the surface of its body, or merely to pass
rapidly before it, to reawake immediately spasms of all
the muscles and general tension of the body similar
to those seen in tetanus.

It is indeed very probably to that same law of the
nervous system that are owing the rare cures obtained
by amputation of the limb (Larrey).

In tetanus (idiopathic) the centre of irritation whence
the nervous action starts is not ascertained; perhaps
from no other than the cord itself, but the rarity even
of that form of tetanus which one may call *essential*,
seems to shew how that affection, still so enigmatical,
is under the influence of the law of reflexion *from* the
cord, since in most cases it starts from some point on
the surface of the body.

In fact we may, in the present state of science, look
on tetanus thus:—

An external lesion being produced, it is exposed to

inflammatory and nervous irritation, more in proportion as it is lacerated or stretched by a foreign body.

That local irritation establishes as a rule a sympathetic and reflex relation with the spinal centre.

It takes on a morbid character, and becomes a complication of the wound in consequence of metastasis of the peripheral irritation to the spinal cord.

Hence—sometimes at the moment when tetanus comes on—the diminution of the discharge from the wound.

The nervous susceptibility of the subject, exalted by fatigue and disease, has, as we have shewn, the effect of accelerating all nervous sympathies, especially reflex action ; it predisposes thus to the tetanic reflexion, and supports it after its onset. For the same reason that the peripheral point of irritation is the starting-point of the sympathetic change, the suitable treatment of this external point can react favourably on the centre. In lessening the stimulus at the periphery, the reflex transmission can be blunted, and the central point proportionately diffused. Unhappily this last is but rarely obtained. That theoretical explanation is founded, as one sees, on ascertained laws of the physiology of the nervous system, and is not without a foundation on experiment.

Our experiences of the treatment of tetanus are not very encouraging. One single cure seems to be referred to full doses of tartar emetic. In others, bleeding and cupping, full doses of opium, camphor, belladonna, potass, baths, etc., have not appeared to have had any decided effect. We cannot speak of the application of cold water, mercurial frictions, moxas, and other recommended derivatives. All narcotics and excitants have been successively tried and found inefficacious. We have noted above the discouraging conclusions of Baron Larrey. Modern authors are not more sanguine.

	Cases.	Deaths.
At the Glasgow hospital .	50	44
St. George's, 1840 to 1854	18	16

Dr. Pollock, in a paper read to the Medico-Chirurgical Society of London, speaks of 8 cases, 1 cure; all were traumatic. However, we shall note at the end of the chapter some cures obtained in the worst conditions.

The remedies employed in the cases of cure marked in the English records, are—

> Extract of Belladonna.
> Indian hemp.
> Quinine.
> Sesquioxide of Iron.
> Dover's powders.

Nicotine, without curing, has had a marked effect on the spasms and general distress.

Bleeding has only appeared to have a passing effect, and we do not venture to recommend it so confidently as other authors. Tetanus not being a true inflammation, losses of blood do not seem likely to have any power, so far as they influence the vascular activity

which accompanies it. In one case of idiopathic tetanus, bleeding had no effect; in another case, the bleeding, which was abundant, appeared to coincide with an
evident exaltation of the nervous restlessness. A
tetanic contraction, acutely painful, with symptoms of
acute fever, fixed on the sphincter ani and neighbouring muscles. The patient died in this extraordinary
condition.

If it be true, as we have said, that traumatic tetanus
is composed of three distinct elements—

An external cause,
A sub-inflammatory centre,
An exaltation of reflex sympathy,

it naturally follows that the treatment should be directed
to these three causes. The wound being the seat of
peripheral irritation, it is necessary to remove that
character from it. We would free a nerve from the
ligature, should it have been included in one; free
over-stretched aponeuroses, extract foreign bodies, especially endeavour to bring back the wound to the most
simple condition, by removing all splinters pricking
the tissues; should the discharge cease or the wound
inflame anew, or become more painful, we would set
up suppuration by emollient poultices, or even by
blisters. Larrey's memoir includes some very remarkable cases where the application of the actual cautery
caused the disappearance of the tetanic symptoms, by
bringing back suppuration.

H

In the English hospitals they seem but little occupied with the treatment of the wound. It is true that in the thirteen burns mentioned the injuries were scarcely susceptible of modification. They do not appear to have noticed any change in the appearance of the wound coincident with the tetanus. Often, on the contrary, it came on when the wound had its best appearance and was healed in part.

An heroic remedy was introduced into surgery by Larrey, and which since has acquired a scientific reputation, viz., *amputation of the injured limb.* Observations on the employment of it are very few; I scarcely know of any save those published by Larrey and Dupuytren ; the former takes his stand on some successful cases, to recommend amputation of the wounded limb at the instant symptoms of tetanus shew themselves. He has seen a case of chronic tetanus cured by amputation. Another case is that of a lieutenant of infantry, who received a ball into his left tarsus; on the twenty-first day he had trismus; on the morrow, general tetanus ; the treatment was most energetic ; removal of splinters, opium, camphor, quinine, poultices; but everything failed, and the state of the wounded man seemed desperate. Amputation, as a last resource, was immediately followed by a general remission ; two months after the accession of the tetanus he was well.

In two other cases Larrey confesses to having only obtained a momentary relief ; once amputation of the leg stopped the spasms for twelve hours ; they reappeared under the influence of a cold night, and the patient sank the third day. On another occasion, a gun-shot wound in the left elbow was followed by intense tetanus; amputation induced a calm for some hours, but the convulsions re-appeared from the cold night air, and death ensued.

As surgical literature contains scarcely anything on this grave subject, I shall recall the fact that Dupuytren performed two amputations under similar circumstances, and that in neither case was the development of tetanus prevented, or the death of the patients. In only one did amputation induce a calm, for *one* day. In conclusion, amputation as a means of curing tetanus, as yet, only numbers two completely successful cases to four unsuccessful.

The second element which assists in the development of tetanus being nervous susceptibility, and especially a centripetal stimulus, one cannot enforce too strongly the importance of withdrawing the patient from all causes of peripheral excitement.

What we have said of the exaltation of reflex sympathy in certain animals, and the ease with which the slightest sensation awakes it, will lead us to keep from our patient the news of the day, noise, conversation, or persons going and coming. We have seen spasm brought on by the wounded hearing musketry, clocks striking, etc., etc.

All that reminds us of similar conditions produced by hydrophobia. I have observed a young girl who sunk in a few days after a fright which placed her in a con-

dition between tetanus and passion (rage). She swallowed
with difficulty, and only when the drink was brought
to her from behind, and in almost total darkness ; my
arrival, or the sight of any one approaching her, set up
tetanic spasms. The post-mortem offered no explanation. To those negative precautions will be added the
employment of direct calmatives, especially full doses
of opium ; Larrey advises its being associated with
camphor.

The third cause, which we have compared to rheu-
matism, and which determines in the spinal centre a
peculiar inflammatory condition, may be treated locally
by usual means taken from the list of antiphlogistics, sudorifics, and alteratives. I shall confine myself
to mentioning alkaline baths, steam, woollen gar-
ments; with tartarized antimony and opium in large
doses.

But there is a method of treatment on which the
English records furnish us with abundant and valuable
information. I allude to *inhaling chloroform.* Of forty-
three cases twelve used this remedy ; of these, four
were cured.

Nevertheless, the following are the conclusions
which seem to result from numerous experiments :—

1. There is not one case of recovery which the
English authors believe can be attributed directly to
the anæsthetic.

2. Chloroform has caused no inconvenience.

3. It has always obtained a remission, but never a
definite diminution of the spasms, either in frequency

or violence. The intermission has sometimes lasted an hour, and the pain has been diminished.

4. It has the great advantage of allowing the patient to be fed during the state of collapse.

5. In conclusion, its employment has not had all the results which its momentary effects seemed to promise.

In the only case of tetanus observed among the wounded of June 1848, of which we possess a detailed account, chloroform was employed but without further effect than the short intermission I have mentioned. He was a national guard of good constitution, wounded on June 24th at the top of his left thigh; the ball was extracted, and the wound seemed trifling.

30*th*, He complained of a trembling of his thigh when he tried to move, and a similar feeling in his right arm, intense trismus, face injected, with contractions of facial muscles, difficult deglutition, could drink no more from this time.

Chloroform.—The first time this remedy relaxed all his muscles, except the masseters ; they only yielded to the third application. On ceasing to apply it the rigidity returned. On the 25th, 80 *centigrammes* of extract of opium.

July 2*d*, Electricity in induced currents from, at first, metal knobs, then with acupuncture needles through the masseters, and in the sterno-mastoid ; slight remission, no pain, except dyspnœa, then some spasms, followed by death.

Autopsy.—The ball had only wounded muscles, no

nerves ; in the nervous centres was only found a slight injection of the membranes, no excessive quantity of fluid nor softening of either cord or brain.

NOTE.—The author here appends a valuable list of cases, but as they are taken from English records, and the conclusions to be derived from them already stated, I think it better not to give them.   T.

PART THE SECOND.

GUN-SHOT WOUNDS CONSIDERED IN DIFFERENT PARTS OF THE BODY.

——◆——

CHAPTER I.

WOUNDS OF THE LIMBS IN GENERAL.

IT may perhaps excite surprise to see wounds of the extremities occupy the first place, but we must remind our reader that this is not a systematic treatise on military surgery, but merely a series of studies on the questions relating specially to gun-shot wounds, regarding which we have collected a considerable number of observations, and which we have been able to submit both to practical observation and to the comparative study of writers on the subject.

Wounds of the extremities have claim to particular notice, as being of all gun-shot wounds the most interesting to military surgeons, from their variety and the efficacy of their treatment. It is to these especially the questions apply of which we have treated in the first part of this work ; and they have given rise to the grave questions of the comparative value of primary or secondary amputations. Wounds of the extremities are, as may be supposed, much the most numerous. Besides the greater mobility of the legs and arms, which increases the chances of their being hit by a ball, it also helps to swell the number of these wounds that they are scarcely ever fatal during the first few days, and thus there is always time for the wounded to receive hospital attendance, more or less prolonged. In fact, it is not so much to the immediate consequences of these wounds, but to the secondary complications, especially to purulent fever, that patients most frequently succumb. Our notes present a total of 79 wounds of the limbs to 58 of the trunk, that is, a proportion of 4 to 3 ; in the Maison de Santé, 58 to 34, a proportion of 3 to 2 ; likewise, Monsieur Baudens, in 1848, 91 to 41, more than double.

Comminuted fractures of the limbs by gun-shot wounds resemble each other in the manner in which cicatrization of the soft parts, and consolidation of the bone, take place. In this respect they may be considered from one common point of view.

On the other hand, their prognosis differs strikingly ; and in the same manner, as it is correctly stated that amputations increase in importance in proportion to their nearness to the trunk, it must be admitted that gun-shot wounds become aggravated according to the same law.

Besides these resemblances, and these general differences, experience has proved that one ought, in regard to the prognosis, as well as the treatment, to *separate wounds of the thigh from those of the other limbs,* and to consider them apart. This division has impressed itself, more or less, on all surgeons. Fractures of the thigh constitute, by their extreme seriousness, a separate chapter in military surgery. It will be seen, therefore, that it is not easy to speak of the treatment of fractures from gun-shot wounds without keeping in view this division, and the statistics given must necessarily vary from one author to another according to the number of thigh wounds comprehended in his estimate. In order to obtain conclusive results upon the different practical points relating to the category of fractures, it is necessary, in addition, to collect a considerable number of exact observations, taken in the most different circumstances in the history of wars, so as to withdraw as much as possible from general consideration, the fortuitous and accidental causes which may have partially influenced the result. Unhappily it is not always possible to submit questions to an examination so general and so rigorous, and it is often left for us only to compare the assertions of different authors on the points in question, and to see to what extent they agree with each other.

Wounds of the lower limbs are generally more numerous than those of the upper. In our list we find, —lower limbs, 52 ; upper limbs, 27.

The list of the Maison de Santé reckons,—lower limbs, 37 ; upper limbs, 21 ; a proportion of 3 to 2.

Monsieur Baudens reckons at the Val de Grace, in

1848,—lower limbs, 51 ; upper, 40 ; a proportion of 5 to 4.

The principal cause of this difference is evidently the greater bulk of the lower extremities. Another circumstance which may also contribute to it, is that the movements of the arms, bringing them generally into a horizontal position, they present a foreshortened surface to the ball, and are therefore less likely to be hit.

CHAPTER II.

FRACTURES OF THE THIGH.

Treatment of fractures of the Thigh in Paris in 1848.

THE treatment of comminuted fractures, and especially those of the thigh, is one of the most disputed subjects of military surgery. It has become in its turn a sort of scientific battle-field, in which the first medical celebrities have taken part.

We may here observe that in the following pages we shall frequently combine what relates to fractures of the limbs in general, with what especially relates to those of the thigh. The latter are, as I have said, the principal objects of scientific interest to the military surgeon among wounds of the limbs, and the principles he seeks to establish are applied especially to that species of fracture.

Do fractures by gun-shot wounds, and especially those of the thigh, generally admit of a conservative treatment; or do they, sooner or later, require amputation? Which are the cases in which either of these courses should be preferred? When amputa-

tion may be considered inevitable from the first, ought
it to be performed immediately, or is it better to wait
until suppuration takes place ?   Such are the leading
questions on which civil, as well as military surgeons
are so much divided.   Beyond these two general questions, the other points relating to the treatment of
fractures are details of practice.

Our experiences on this subject were collected in
the hospitals of Paris.   We have preserved notes of
a score of cases of fractures of the thigh.

Out of this number, let us say at once three, or four
at most, presented some well-grounded hopes of cure,
the others promised almost certain death.   The prognosis was, in fact, everywhere most unfavourable, and
there was no concealing that the wounded could escape
death only by running the gauntlet of a succession
of most dangerous symptoms—hemorrhage, amputation,
when it has taken place at once, purulent infection to
which thigh wounds have appeared to me more exposed than others, and loss of strength from the suppurative discharge; finally, the well-known danger
of secondary amputation, or, if that has not taken
place, an imperfect union of the femur.

At the period of the Revolution of 1848, French
surgery was essentially conservative.   I do not think
I saw more than three or four primary amputations of
the thigh.   Professor Roux tells of having performed,
in June 1848, 1 primary amputation of the thigh, 2
secondary.   M. Huguier from 5 fractures performed
4 amputations, of which 2 were primary.   M. Jobert

tells us himself — " My comminuted fractures of the thigh have been unhesitatingly submitted to the same treatment as fractures occasioned by any other cause than a ball." Baudens is, however, very favourable to amputation, and performed after the days in June, 5 amputations of the thigh, of which 1 was primary. M. Monod, out of 5 fractures performed 2 amputations, both secondary.

A battle in the centre of a great city where hospitals and all the resources of art abound, enables surgeons to treat the wounded almost as they would do in civil practice.

Thus surgery had in some degree lost its military character, and there was an almost universal desire to amputate as little as possible, and to combine as much as possible all the means offered by genius and science to preserve the fractured limb at any risk. I have not observed one single case of good and substantial union. One of my colleagues, then an *interne* of the hospitals, has confessed to being not more fortunate than myself. He has seen, as I have done, only exhaustion from excessive suppuration, death by purulent absorption, or secondary amputation with its feeble chances. The lists published by the surgeons themselves, almost all speak the same language ; we shall give them all at the end of this chapter.

We give here a resumé of two typical cases :—

Observation 1. A man aged 30 ; bilious temperament ; constitution of average strength. Ball entered in front in the middle of the thigh ; comminuted fracture of the femur, put upon an inclined plane; after some days of excessive suppuration the patient begins to get exhausted; the least movement makes him scream violently; the ends of the bones are with difficulty kept in position; long and deep counter-openings in the region of the trochanter; extraction of a splinter.

12th July. State middling; little sleep; discharge of pus abundant, but healthy; suffering extreme. *17th,* Second counter-opening lower than the first; extraction of a splinter; excessive purulent discharges. Cure still more doubtful. *26th,* Same state. Extraction of four or five strong splinters; the pieces become more and more movable; the adaptation more difficult; weakness progressive. Death.

Observation 2. *25th June,* Single wound in the fore part of the right thigh, at five centimetres only above the groin; clamminess, pain; incision and extraction of a regulation ball and of several large splinters, but several others are felt in front. *17th July,* Extraction of a new splinter, and the following day of four others about six centimetres in length. General state good enough, except a little diarrhœa. *24th,* Splint of Scultatus applied. *11th August,* Excessive diarrhœa; very painful sloughing on the sacrum; exhaustion. Splint of plaster with slits at the edges of the wounds. Fresh abscesses appear on the back, at the groin, on the calf of the leg. The sufferer gradually and quietly expired.

A satisfactory progress of several days, as we have

sometimes observed, is unhappily not always a proof of certain amendment. The notes we have collected relating to fractures of the thigh, often speak of the satisfactory condition of the patient—of the suitable application of the splints, but soon after, more serious symptoms have suddenly supervened, and the patient has sunk.

On the other hand, it must be acknowledged, that the skilful surgeons to whom are entrusted the duties of the hospitals, do not deceive themselves as to the slight chance of success which this treatment offers, however carefully it may be directed. Seeing themselves in possession of all the requisite resources for the preservation of limbs which would formerly have been given up at once to immediate amputation, and hoping, by means of care and watchful attention, to save some, they have all naturally been induced to use the expectant method. Every surgeon, either in the hope of making a step in advance in that difficult surgical question, or from simple humanity, would have acted in the same manner.

Unhappily these efforts were not in general crowned with success, and academic meetings, at which these new experiments received the sanction of scientific debate, have corroborated this opinion. On the other hand, it is beyond doubt, that if the numerous gunshot wounds which we have had occasion to observe simultaneously had been treated after an ordinary battle-field, and in the conditions of camp life, the same surgeons would have been driven by force of circumstances to perform a greater number of amputations. The conservative method, which has predominated, is then in a great measure due, it would appear, to the circumstances in which they have been placed.

To satisfy ourselves of that, it is sufficient to remark that the greater number of surgeons in Paris, after the fruitless attempts which have been made to preserve broken thighs, have pronounced in favour of amputation, viz., Roux, Jobert, Bégin, Huguier.

Can fractures of the thigh be cured without amputation, and in what proportion ?

Let us now examine the two general questions which we said at the commencement lay at the root of our subject :—1st, *Does a thigh fractured by a ball require amputation or not ?* 2d, *Amputation, once deemed necessary, ought it to be performed immediately ?*

It is difficult, in reading the different authors on this subject, to distinguish clearly between the two questions I have just pointed out. The alternative between amputation done on the spot and amputation delayed, is often identified with the choice to be made between an attempt to preserve the limb and its immediate amputation. From whence comes this confusion ? Evidently from this, that the term secondary amputation has not always been taken in the same sense. This term may in the first place signify :—In the case of a fracture for which amputation is judged indispensable from the first, intentionally to delay that operation, in the idea that if performed later, it will offer better chances of success. It may signify in the second place :—In a fracture of the thigh, the union of which is not absolutely impossible, to attempt its preservation, and only to amputate later when the attempt has miscarried to the extent of endangering the life of the patient. In the face of this confusion of the two questions, it is not always possible to consult authors separately upon each of them. It will, consequently, not create astonish ment if they are sometimes found together in the con siderations we shall have to offer.

Let us ask, first, if fracture of the thigh by gunshot wound is capable of union. We might have doubted it from our own observations. Nevertheless, cases of union, though rare, are positive facts, and it may be interesting to cite some well-authenticated examples. Let us speak first of the results obtained on this point among the wounded of the Revolution.

1st Case, Royal Maison de Santé. — Comminuted fracture of the right thigh, wound of entrance at the internal and middle part, wound of exit at the posterior and middle part. Several considerable splinters were removed, which, being put together, formed almost the complete thickness of the femur. In October the consolidation was scarcely complete.

2d Case.—25th June. N. has not experienced serious general injuries. No splinters extracted dur-

I

ing the first days. In front, and behind the upper portion of the right thigh, the entrance wound was lengthened vertically by an opening. Above and underneath, the wound was of a grey tint, through which could be felt the end of an immovable splinter, which appeared to have a transverse position. All the thigh inflamed; very painful to the touch; general health pretty good. At the beginning of October a new abscess formed in the inner part. From that time the pain and swelling diminished, suppuration became less, the wounds closed, with the exception of the upper internal one, which had gray swollen edges. The whole thigh can be raised at once; there was the commencement, distinct, of consolidation. Convalescent.

3d Case.—A National Guard was received on the 15th June at the Surgical Clinic of the Faculty (attended by M. Giraldès). The ball broke the left thigh, entering in the middle from outside, and going out behind at the same level, from whence it was extracted by an incision. General state, very good. The limb was placed afterwards in the apparatus of Scultetus. After the 30th June the thigh was placed in a concave splint of gutta percha, which kept it from moving. *10th July.*—The patient moved to a mechanical bed. Goes on well. The wound in front is almost closed; behind, there are two wounds also nearly closed; but there is a third wound from which pus flows, and a mass of small fragments were found in it. Some had been extracted on the 26th July. There was a point within the fracture which gave exit to matter. General condition good; appetite satisfactory. No rigors, but evident emaciation. No diarrhœa. *25th October.*—Very satisfactory; patient eats and sleeps well; his thigh has consolidated, but with an enormous callus and decided shortening. The patient can easily turn in his bed. There are still fragments perceptible in the callus, but they cannot be extracted. A fruitless attempt at extraction was made on the 28th June.

4th Case.—M. Amussat gives a detailed account of a comminuted fracture of the left femur, in the middle, from a gun-shot wound received when standing, in 1848. A double wound; great hemorrhage; thigh enormously swollen, deformed, and shortened. In spite of the horrible pain, and contrary to the wishes of the patient, they tried to preserve the limb; on the inclined plane, with continued irrigations of tepid water. In the course of treatment they had to encounter successively phlegmonous erysipelas, and sloughing of the cellular tissue. Enlargement of the wound for the extraction of bullet and splinters. Amputation frequently proposed, and rejected; at last, at the end of the sixth month, the patient was out of danger, with the limb united, but an unsightly callus, and considerable shortening. He got up and took exercise in a mechanical chair. Baudens tells of having attempted to preserve the thighs of 20 individuals, and having cured 2 with deformed limbs. Malgaigne tells of 2 cures out of 4 fractures of the thigh, treated without amputation. Jobert tells of 2 comminuted fractures of the thigh cured by secondary callus. One of my colleagues related to me the case of one of the wounded at the battle of Novara, who was cured of a fracture of the femur, and who had been seen by him several years later.

J. Bell, in his enumeration of cures of broken limbs without amputation, finds only two authentic instances after fracture of the thigh.

Dr. Stromeyer cites four complete cures of broken femurs, of which two were without shortening of the limb.   They had been wounded at Fridericia, and treated in the same town by Danish surgeons.

M. Quesnoy says that he saw in the hospitals at Constantinople several cases of perfect consolidation of the femur with shortening.

Let us conclude from these facts, which certainly do not include all that surgery possesses on this subject, that consolidation has been observed by almost all the surgeons of the army; all know more or less of these cases, the authenticity of which can not be called in question.

Let us add, however, that almost always there is the chance of very considerable shortening of the limb, and of an unseemly and voluminous callus, after a very painful course of treatment which has lasted sometimes more than a year.

It is difficult, unhappily, in these cases, already few enough, to decide exactly what is the use to which the cured can put his limb.   Accurate reports on this subject are rarely furnished; the cases where the patient has been able to walk freely and without any obstacle, are, we must acknowledge, among the number of surgical curiosities.

To resolve in a general way the question whether fractures of the thigh make better recoveries after amputation, or attempts at preservation, it would be not only necessary to know the results obtained by each surgeon, but also to have an analysis of all the cases, with notes of their different forms, of the seat of the wound, of the state of health of the patient, of the mode of treatment, together with the hygienic circumstances with which the patient was surrounded at the time. Unfortuately, in general we can but get a summary, founded sometimes on a mere approximation. If we tabulate the cases we possess of fractures of the thigh treated *without* amputation, the following is the result :—

	Cases.	Cures.	Deaths.
Baudens (1848) ; upper part of the femur, at the same time fracture of the infra-orbital region	1	0	1
—— In middle part	1	0	1
—— Great trochanter	1	1	0
Malgaigne (1848)	4	2	2
Gosselin (do.)	3	1	2
Dupuytren (1830); at the Hotel Dieu, excluding fractures of knee and hip joints	13	5	8
Jobert (1848)	6	2	4
	29	11	18

This gives the proportion of recoveries as 11 to 29, that is to say 38 per cent.

We have been fortunate enough to procure, by the kindness of Dr. Monod, head surgeon to the Imperial Maison de Santé, the results contained in a manuscript work founded specially on observations collected in that hospital.

This memoir, which we shall have occasion to make use of again, indicates for fractures, in which amputation was *not* performed, the following proportion :—48 cases collected in different hospitals, of whom 17 recovered, that is to say, 35 per cent. If now we glance at the *table of amputations of the thigh* from various authors, we find :—

	Cases.	Deaths.
M. Roux, 1830, primary	1	0
„ „ consecutive	3	3
„ 1848, primary	1	1
„ „ secondary	3	2
Dupuytren, 1830, primary	5	3
„ „ secondary	4	4
Huguier „ primary	2	0
„ June, secondary	2	0
„ January, secondary	5	0
Baudens, primary .	2	2
„ secondary .	4	4
	32	19

	Cases.	Cures.	Deaths.
Total primary . .	11	5	6
„ secondary . .	21	8	13
Total	32	13	19

32 amputated, 13 cures, gives a rate of recovery of 43 per cent. A comparison of these two tables would thus shew a marked superiority on the side of amputation.

Do fractures of the thigh require primary or secondary amputation ?

Before pronouncing an opinion on amputation, we must compare primary with secondary amputations. The two questions we now examine have not in general been separated by writers on the subject, and ought, therefore, to be combined in our present study, although we separate them in practice. Ought the amputation of a limb, and especially of the thigh, when fractured by a ball, to take place immediately, or would it be better to delay it until a more or less advanced period of suppuration ?

I shall first mention succinctly *the opinions of the most distinguished surgeons* upon this question.

Duchêne in his *Treatise on the General and Particular Cure of Arquebusades*, in 1625, appears to have been the first who recommended immediate amputation in wounds of the extremities. " It is necessary," he says, that the operation should take place before inflammation and other general symptoms supervene." In 1737, Ledran pronounced in the same manner for primary amputation in preference to secondary. In 1745, after the battle of Fontenoy, the Royal Academy of Surgery of France offered a prize for the best essay

on primary amputation. It was, as every one knows,
Faure who carried off the prize. Though the result of
his researches led him to condemn primary amputation,
he nevertheless acknowledged the necessity of it in
some cases which he pointed ont.

The famous Bilguer, surgeon-general of the armies
of the king of Prussia in 1764, a native of Coire, in
the Grisons, published at that time a dissertation—
" *De Membrorum Amputatione rarissime adminis-
tranda aut quasi abroganda.*" He prohibited surgeons from practising amputation, which was from that
time abandoned throughout the Prussian army. More
than 6000 wounded were thus treated without ablation
of the members. It is not surprising that out of so
great a number some marvellous cures should have
been obtained. It is curious, in opposition to this
method, to look at that followed a few years after, in
1785, by another surgeon-in-chief of the Prussian army,
J. L. Schmucker. He gives a detailed enumeration of
the wounds which required immediate amputation, and
considers it indispensable, particularly in fractures of
the thigh above the middle part. He tells of having
performed a great number with complete success.

John Hunter was, as is known, a most decided
partizan of delayed amputation. He calls immediate
amputation very bad practice. He thought that the
sudden loss of a limb, sustained by a body in full
health, is a great risk to the constitution and life of the
patient. One may almost always, he says, wait until
inflammation is established ; if the wounded man does

not stand that, it is probable that he would not have
borne immediate amputation ; whilst, if he bears it,
he has also a good chance from secondary amputation.
Building upon the good results of amputations performed for long and debilitating maladies, he concludes
from them that this operation will be better borne
when, from suppuration, the patient has got in some
degree into the condition of one suffering from a
chronic affection. The celebrated English naturalist
is thus, as we see, directly opposed to modern military
surgeons, and especially to the authority of Larrey.
But we do not fear to say, that great naturalist enjoys
a reputation so universally acknowledged in the natural
sciences properly so called, that we may be permitted
to throw some doubt on his competency in a question
of purely practical surgery. He remained only a short
time in the army, at the siege of Belle Isle in 1761.
In 1762 he went to Portugal, where he gave himself
less to the study of military surgery than to researches
in natural history. In 1792 Baron Percy, surgeon-in-
chief of the French army, declared himself the absolute
enemy of primary amputation. Finally, the most decisive opinion against all temporizing is Baron Larrey's,
an eminently practical genius, and brought up in the
school of battles. The opinion of this great military
surgeon upon these grave questions is real authority.
It is in war, in fact, and in a great war, that one learns
to take account of material possibilities—necessities of
the first order. It is then, consequently, that the
prestige disappears of apparently the best founded

physiological theories. Thus, we see modern military surgeons, for the most part, range themselves on the side of operations at the commencement; such are— Bégin, Baudens, Dupuytren, etc. "We ought henceforth," says Larrey, " to have but one opinion on the subject. It is after having directed for twenty years the sanatory service of the army, that I come now to lecture in the Academy, and to solve definitively that great question which I look upon as the most important in military surgery." Guthrie, without pronouncing absolutely for immediate amputation, says, that when it is indicated it must not be delayed longer than from one to three hours after the commencement.

But let us hasten to arrive at *Modern Surgery*, and those recent opinions matured by the experience of 1848 and the Crimean campaign. If the question of amputation, compared with the attempt to preserve the limb, may be considered as still presenting differences of opinion, on the other side, modern surgeons are all agreed in giving to primary amputation the preference to secondary. None of the surgeons who directed the hospitals of Paris in 1848, delayed an amputation which was judged to be indispensable from the beginning; and when they performed secondary amputation, it was always because they were driven to it by limbs which they hoped to have been able to preserve, and which, after a fruitless attempt, were found to endanger the life of the patient.

It is probable, then, that if Faure's work in favour of secondary amputations appeared in the present day, he would no longer be successful. In the same manner, the opinion of John Hunter is no longer mentioned in France on this point, except as a tradition. Let us remark also, that if some are opposed to immediate amputation, this ought not, for the most part, to be interpreted as giving a preference to secondary amputation, but as an opinion favourable to attempts at preservation. Professor Roux expressed himself before the Academy in August as follows : " I shall lay down, as a principle, that immediate amputation of the limbs ought to be performed more frequently than it is in general ; and I shall not reproach military surgeons with sacrificing too often, on the field of battle, limbs, the preservation of which it might, strictly speaking, have been possible to attempt. In every case, I fear less excess on this side than on the other."

M. Malgaigne is the person among modern surgeons most strongly opposed to amputations in general, and in particular to immediate amputations. It would be too long to submit the statistics which he has furnished to examination.

M. Velpeau is little in favour of amputation :—" I am disposed to perform the smallest possible number of immediate amputations, and to try to preserve the limb in the greater number of cases. On the whole, the older I grow in practice, the less am I an advocate of immediate amputation. I amputated more in 1830 than in 1848, and in June less than in February." M. Huguier practised in 1848 primary amputation only for fractures accompanied with great destruction of parts ;

the results of his 15 amputations have been astonishingly favourable, since he had 15 cures.

M. Jobert pronounces in favour of the attempt to preserve limbs where the flesh has not been too much crushed, even when numerous wounds exist, and numerous splinters which pierce the tissues; but, when ablation of the limb is judged necessary, he gives the preference to immediate amputation, as superior to the secondary operation, provided that they wait about twenty-four hours, so as to allow the patient time to recover his senses.

Professor Stromeyer, surgeon-in-chief of the army of Schleswig, does not pronounce a decided opinion. He thinks that union ought to be attempted when the ball appears to have produced only a simple fracture. The campaign of 1849 has offered numerous cases of union of fractures of all kinds. He thinks that the question is not yet decided, and that its solution on one side or the other depends on the facility one may have for maintaining the limb in an immovable position, and not being forced to transport it beyond a short distance.

Comparative Table of Primary and Secondary Amputations.

	PRIMARY.			SECONDARY.		
	Total Number	Cures.	Dths.	Total Number	Cures.	Dths.
Guthrie, after the battle of Toulouse	47	38	9	51	30	21
Ibid., attack on New Orleans . .	45	38	7	7	2	5
Dupuytren . .	7	2	5			
Baron Larrey . .	13	11	2			
Roux, 1830 . .	10	7	3	4	4	
Ibid., February 1848	1	1	...	3	3	
Ibid., June . .	11	5	6	6	2	4
Ibid., says elsewhere in adding all the figures . .	23	13	10	9	6	3
Baudens . . .	14	11	3	6	...	6
H. Larrey, 1830. Gros Caillou .	6	3	3	11	5	6
Ibid., siege of Antwerp . . .	54	45	9	10	5	5
Laroche Lyon, 1834	19	6	13			
Returned from the Maison de Santé 1848. Dr. Monod	13	12	1	2	...	2
	263	192	71	109	57	52
Proportion of cures.		73			52	

After this result we cannot hesitate to give the preference to *primary* amputation.

If from this general comparison we pass to that of *amputations of the thigh*, we find, according to the table given at page 118, for immediate amputations, a proportion of 45 per cent, whilst secondary operations furnish only 38 per cent. Here, also, the advantage rests with immediate amputation. L'Hotel des Invalides contained, in April 1851, out of a total of 3167 men, 102 cases of amputations of the thigh, 43 of which had been performed on the day, or the following day, and 69 before the 30th day. One may conclude from this proportion that immediate amputation was, before 1851, in favour among military surgeons. As to the proportionate mortality, that cannot be deduced from these figures, which do not shew out of how many amputations of the thigh there were these 102 cures. Besides, it must be observed that these ablations were not all performed for fractures of the femur, which may have had an influence on the result.

The following observation, taken among many others, is an example of the grave aspect an amputation may suddenly assume when performed after the patient is already weakened by prolonged suppuration:—A ball entered the middle of the right thigh on the outer side, and went out on the inside at the same height, fracturing the femur; splint of Scultetus was applied for some days; a sanguineous discharge, and gases formed in the wound; amputation was resolved upon; the 17th day, chloroform; although the patient had not lost more blood than usual, his lips were colourless, and he sank the same afternoon, with the symptoms one remarks after excessive hemorrhage. The autopsy shewed a comminuted fracture of the femur, bathed in a large pool of suppuration.

Let us now compare the results of our table with those furnished by the manuscript work which we have already mentioned, and which we owe to Dr. G. Monod of Paris. The statement it contains is founded upon the comparison of 256 observations of comminuted fractures, taken in the practice of nine surgeons in Paris. It will be observed that these statements differ from ours, in that they have been drawn up only from experiments made upon the wounded of June, whilst ours are taken from all the sources placed at our disposal, without regard to time or place.

Here are these results :—

	Cases.	Deaths.	Proportion of Cures p. 100.
256 Comminuted fractures of the limbs in general . .	158	98	62
45 Immediate amputations for comminuted fractures in general	29	16	64
178 Treated without immediate amputation (35 of which secondary amputations) .	109	69	64
35 Immediate amputations of other parts than the femur and the knee	26	0	74

	Cures.	Deaths.	Proportion of Cures p. 100.
107 Not amputated, with other members than the femur and knee	86	0	83
6 Immediate amputations for fractures of the femur and knee	2	4	33
48 Fractures of the femur and the knee not amputated .	17	31	35

If now we sum up the conclusions which seem to result from our two last tables we shall find :—

1st, That, taken altogether, comminuted fractures of the four limbs heal better after immediate amputation than the cases treated without immediate amputation (2d Table). This result must necessarily vary according to the number of fractures of the *thigh* included in the calculation.

2d, Primary amputations generally afford better results than secondary amputations (2d Table).

If we divide the total of the fractures into two parts—(*a*) those which concern other members than the femur and the knee, (*b*) those relating to the latter, we find,—

3d, That fractures of the leg, foot, and upper limbs, have more cured, among the non-amputated, in the proportion of 83 to 74. This conclusion does not imply, as may be supposed, that amputation should be counter-indicated, but only that those wounds may be more frequently cured without that operation, and that amputation being generally practised upon the most serious cases, naturally presents a smaller number of cures.

4th, Fractures of the thigh present the least reparative power of all—2d Table.

5th, A list of 6 immediate amputations is too limited to admit of a decisive comparison with 48 cases of fracture not amputated. The difference of 33 per cent and 35 per cent has also little force, and does not in consequence sufficiently sustain the superiority of non-amputation. The difference we have found in favour of amputation seems more decisive (a proportion of 13 to 11).

6th, Primary amputations of the thigh present better results than secondary, in the proportion of 45 to 38.

Immediate amputation of the thigh thus comes out very advantageously from this comparison of figures. But before expressing ourselves more positively, let us see what is the result of *experience acquired in the Crimea* on this point. These are the terms in which a work on this subject commences, entitled, " On the Treatment of Wounds from Projectiles of War on the Wounded of Alma and Inkermann. By M. Valette, Physician-major of the First Class." " The deplorable result of the greater number of secondary amputations in the service, with which I have been charged, affords one more proof of the superiority of primary amputations, and the necessity of using freely in the ambulances that kind of practice which, in doubt-

K

ful cases, is more disposed to remove at once the wounded limb than wait for secondary amputation."

M. Quesnoy expresses exactly the same opinion, saying, " In all cases, in the Crimea, amputations were performed as quickly as possible, for, it must be observed, that besides the advantages experience now accords to immediate amputation, the patients are much more disposed to submit to amputation the day they are wounded than the following day, or later."

Out of 30 immediate amputations, 23 of which were important operations, this writer reckons 18 recoveries, a proportion of 60 per cent; our second table gave 64 per cent. From great secondary amputations the number of deaths has been in the frightful proportion of 7 out of 9. The proportion of 60 per cent is still more satisfactory when one thinks that in this case it is not, as in 1848, a street war, when men were struck down in complete health and vigour, taken to the first hospitals in the world, and entrusted to the care of the élite of the profession. Unfortunately, correct statements of the specific results of amputation of the thigh are very few. Of 4 cases, 2 in the lower half, 2 in the upper; the two last sunk. Dr. Scoutetten, in his abridgment of medico-chirurgical observations made with the army of the East, concludes, from two comparative tables, that primary operations have been much more frequently followed by success.

Experiments tried in the Crimean war, with the view of preserving fractured limbs in general, appear to have been unfortunate, and of course, still more so in fractures of the thigh, the most serious of all. The improvements that have taken place since the last wars, whether it be in the treatment, or the means of transport and of dressing, encouraged surgeons more frequently to attempt the preservation of fractured limbs. An author, who played an active part in these events, expresses himself in the following manner :— "This fatality of wounds has been aggravated in no slight degree by the zeal of the surgeons in wishing to preserve injured limbs. Full of the promises and teachings of the schools, they tried to persuade themselves that wounds apparently slight could not long defy intelligent and assiduous endeavours made to save a limb. Many lives paid the penalty of this disastrous experiment, and confirmed the precepts of the masters of camp surgery."

It has been demonstrated anew, that any attempt to preserve a limb in the conditions mentioned endangers the life of the wounded, and that the chances which follow secondary amputation never restore a balance in favour of the patient. This want of success was due, besides the useless attempts of which I have spoken—1st, To the deplorable state of health of the soldiers ; 2dly, To the new nature of the projectiles, and especially of the conical balls, which produce in the bone a much greater number of splinters than the ordinary bullet. Although our experience upon the two questions which we have discussed—1st, Amputation compared with attempts at preservation; 2d, Immediate amputation as compared with secondary—is confined to observations collected in the hospitals—let us finish by some observations which embody our own opinion on the subject.

The authentic statistics we have given, and the numerous failures we have observed in 1848, owing to the attempts at preservation of limbs, have convinced us, that if the first duty of the surgeon is to save the life of his patient, he ought not to let himself be led away by the deceitful hope of sparing him a deformity. His duty, especially in war, is to amputate immediately, every time there is a comminuted fracture of the femur, if the wounded man is not reduced by hemorrhage, or any other wound, to a perfectly hopeless state.

I except only fractures in the upper portions of the femur, which would necessitate either amputation closely approximating to articulation, or amputation at the hip-joint—two operations, almost always fatal, and which on this account, perhaps, still leave the advantage in favour of attempts to preserve the limb.

If I required, after the great authorities referred to at the beginning, to depend upon a name which has scarcely its superior in French surgery, I should quote these memorable words of Dupuytren—"What, then, is the rule of conduct we ought to follow? to amputate from the commencement; not to let oneself be carried away by too fragile hopes. Military surgeons were formerly accused of being too ready to amputate, I have often said, and I now repeat it for the last time—my opinion upon this point is unshaken; in complicated fractures from gun-shot wounds, in rejecting amputation more individuals are sacrificed than limbs are saved."

With regard to the exception I made, relative to immediate amputation of the thigh in the upper part, I am happy to see it confirmed by experience in the Crimean campaign. Thus, M. Valette says, that of four amputations of the thigh, two in the lower half were cured, two others performed in the upper part failed. He tells also of having, at Canlidje, performed three amputations in the upper part of the thigh, all followed by death. Dr. Legouest, physician-major, professor in the Imperial Military School of Val-de-Grace, in his essay upon coxo-femoral disarticulation, grounds upon experiments in the eastern campaign in cases of wounds in the region of the trochanter, and the neck of the femur, in order to dissuade from disarticulation, and to recommend in preference the preservation of the limb. He does not pronounce absolutely against amputation *near* the joint, as he does in regard to disarticulation, but he prefers attempts at reunion. Amputation of the upper part of the thigh has presented, more than any other ablation, cases of death from exhaustion, not from hemorrhage, but from nervous depression owing to the large surface of flesh laid bare by the knife.

As we have the opportunity, let us conclude by some words on experiments furnished by the Crimea, in regard to coxo-femoral disarticulation. After the battles of Alma and Inkermann, that great ablation was performed thirteen times ; all were followed by death.

Dr. Legouest has, however, been so fortunate as to attain almost complete success in one case. It was that of a Russian prisoner, aged 30, whose left thigh was penetrated by a ball close to the articulation, at the battle of the Alma. Thirteen days after, they performed exarticulation. He recovered, so far as to be able to get up. For more than a month they did nothing more than nourish him well. Then he had the misfortune to fall upon the stump, which brought back the inflammation, and caused his death four months after the operation. This brilliant success, far from increasing the confidence of the author quoted, in exarticulation, makes him completely reject it, for all the other cases were much more rapidly fatal, whilst out of 10 fractures of the coxo-femoral articulation not amputated, 3 were cured with preservation of the limb.

The Hotel des Invalides presented, in April 1851, out of 3167 men, one single case of coxo-femoral disarticulation, which was performed three years after the accident.

In collecting all the *primary exarticulations* of the thigh mentioned by authors, we find 30 cases, but also 30 deaths. Out of 11 cases of secondary exarticulation, 3 successful ones are recorded; the case the Hotel des Invalides is also included in this list. It appears, then, that there is rather an advantage in waiting, as the chances are greater both for preservation of the limb in the first instance, and also for amputation, should that be requisite afterwards.

There still remains *resection of the upper extremity*

of the femur, which already reckons a third successful out of a dozen cases.   This operation does not appear to have been tried in the Crimea.

4. ON THE DANGERS OF AMPUTATION.

After having thus pronounced in favour of amputation and immediate amputation, founded on experience and the opinion of the greater number of military surgeons, it is nevertheless of importance that, while judging attempts at preservation of the limb most dangerous to the life of the patient, we do not exaggerate the safety of great amputations.

Two great dangers are consequent on such ablations. That of the sudden withdrawal of a considerable portion of the body, from whence necessarily results an entire disturbance of the organisation, especially in the circulatory system.   It is undoubtedly with a view to obviate this great disadvantage, that John Hunter opposes immediate amputation, and postpones it to the time when the body has been weakened by suppuration.   The organic revolution (répercussion) which in a body in full health produces a genuine plethora, on an enfeebled one has no more effect than to favour its nutrition. Hence the rapid fattening often seen of those who undergo amputation after a prolonged suppuration.

But there is another danger which has been especially observed in the Crimean war, I mean, *the laying bare of a large extent of soft parts.*   Some of the wounded amputated at the thigh, on the field, having only lost a small quantity of blood, and presenting no other bad symptom than a certain nervous insensibility, have sunk a few hours after a well-performed operation and properly applied dressing.

Nothing could explain their death except the extreme nervous depression caused by the baring of a great thickness of tissue.   It would appear as if the vital principle escapes from the wide opening in the organism.   Hence the importance of closing the wound following an amputation as soon as possible, and as *hermetically* as possible.   Perhaps, when explaining this enigmatical statement, we should take into consideration the loss of caloric.   The Crimean campaign offered some sad examples.   For instance, a sergeant is brought to the ambulance with his thigh broken and splintered; they perform amputation. He loses scarcely any blood, and had no bleeding previously.   He is carefully dressed and laid in a litter.   In a few seconds afterwards he falls into a fatal syncope.

These inexplicable deaths have been observed especially in subjects who had been overcome by stupor, and who had lost a good deal of caloric by their exposure on cold and damp ground.   One can conceive, after that, the importance of a maxim generally laid down, only to amputate after the patient has had time to recover from the nervous shock.

Both in Paris and the Crimea the danger has been acknowledged of amputating before the patient has revived from that comatose condition in which he sometimes remains for several hours.

The record of the Hotel des Invalides, published in 1854 by M. Hutin, head physician of that establishment, led to very unexpected results as to the reunion of fractures of the thigh. From 1847 to 1853 sixty-three cases were received into the house which were not amputated, 21 amputated. Among the first there were —

18 Below ⎫
28 In   ⎬ middle third.
17 Above ⎭

Similarly, of 25 fractures of the middle of the thigh, there were 5 amputated for 20 cured without amputation. Of 35 fractures below the middle there were 16 amputated for 19 not operated on. Above the middle not one was amputated out of 24.

It follows, therefore, from these statements, that for fractures of the middle of the thigh there are four times as many " invalides" alive among the non-amputated, and for fractures above the middle there are none which have been amputated, whilst there are 24 who have got well without amputation. It is evident that these tables do not explain the proportions of cure between amputation and preservation of the thigh ; for that, it would be necessary to know the absolute total of the fractures treated by either method ; nevertheless they have their value, as shewing the curability of fractures of the thigh. See, moreover, what condition those with reunited femurs were in, according to M. Hutin's observations. All have shortening, and some deformity of the thigh, but none suffer from *habitual* pains, as most wounded men do.   They suffer less often than some of the amputated do, and then their pains are neither longer nor more violent.   There is not, for instance, constant swelling, abscess or fistulous openings ; fatigue leads sometimes to inflammatory swelling (tension) ; but during the eight years M. Hutin has been there, till 1854, no one has applied at the infirmary for these symptoms ; and, to speak of them as a whole, it is more than twenty years since their limbs were the seat of any symptoms connected with their wounds.

Moreover, preparations shew that there was good union, and that the bones were as healthy as possible. Indeed, those of the wounded who were alive in 1854 affirmed that they preferred to a wooden limb the one that had served them so many years, and the shortnings of which they could remedy by high-soled shoes.

We owe to scientific impartiality, not to pass over in silence a source of information as important, though it would appear a contradiction to our opinion as expressed above, which shews us that, in a question of experiment and practical observation, it is seldom possible to find all opinions ranked on the same side.

Thus M. Quesnoy, who fully acknowledges the immense superiority of primary over secondary amputation, is not quite so satisfied as other army surgeons about amputation in general.   "During the expedition of Zatcha," says he, "we were induced to preserve

many comminuted fractures of the thigh, since we
saw at Constantinople a great number of favourable
examples of the same injury cured without amputation.
In the face of these facts it appears to us possible to
restrict the necessity of amputating every *comminuted*
fracture of the thigh, and to say that amputation is only
necessary when the bone has been broken into *nume-*
rous splinters. If the latter be not numerous, or if the
ball has traversed the bone in the neighbourhood of its
spongy parts, one may attempt the preservation of the
limb even in war time." Besides, he says, "we had
occasion to see in the hospitals of Constantinople
patients with comminuted fractures of the thigh, which
had terminated in the removal of splinters, perfectly
cured, with sometimes considerable shortening it is
true, but better a short thigh than a wooden leg!
In the course of February I counted 15 cases in the
hospitals of Pera, Gulhané, and the University, recovered
from fracture of the thigh, after several months' stay in
these hospitals."

The facts which we have successively enumerated
have a value we are bound to recognise, and which,
as we understand the matter, have become in the hands
of some authors, M. Malgaigne, for instance, an argu-
ment against surgeons favourable to amputation. These
facts shew in all cases that there are in amputation dangers which experience alone can appreciate, and
about which we must not deceive ourselves.

On the other hand, they are not sufficient to shake
us in the general opinion we hold.   We acknowledge

that in holding up amputation as a *rule*, we would lose some patients who might have been saved by attempts at reunion, or deprive others of limbs they might have kept.

The almost impossibility for the surgeon to determine in a narrow wound the extent of fracture of the bone ; the want of time often during action to interrogate the wounded man ; to reflect; to consult his colleagues, and make up his own mind; the difficulty of suitable transport, and of attentions continued to the end ; all these circumstances, joined to the extreme mortality of non-amputated fractures observed in 1848, oblige us to allow for field surgery a hardly uniform line of conduct.

5. INDICATIONS FOR PRIMARY AMPUTATION.

We have no new matter to add to what has been already said by authors on this subject.   The indications for amputation vary with one surgeon, or another, according to his ideas upon the possibility of preserving the limb.   If it be admitted that, as a general rule, fracture of the femur by a ball renders amputation necessary, one will be less scrupulous about determining the special complications which demand it.   As to the other limbs, the chances of recovery being much greater, the points of detail laid down by authors are applied to them.

The following are the conditions in which amputation is considered necessary :—

1. When a limb has been carried away by a bullet. Larrey cites a series of observations, proving that when under these circumstances they delayed equalising the wound, they lost the patient.

2. When the bones are broken and the soft parts deeply lacerated.

3. When a limb has lost a great portion of its soft parts without fracture of the bone.

4. When the limb is broken and only communicates with the body by the part where the main artery lies.

5. A ball towards the end of its course has broken a limb without rupturing the integument.

6. A ball has smashed a joint, or the ball has remained within one.

7. A long bone is bared to great extent.

8. An opening into a large joint by a cutting instrument, etc.

This enumeration is far from being complete, for it is difficult to foresee all the cases which result from the varied chances of war. But what appears to us more important is to connect that surgical dogmatism with one common physiological principle. What are the parts of a limb which, by their destruction, entail, sooner or later, 'the loss of a limb? Some experimental physiological details on this question would have the advantage of simplifying the subject, but, so far as I know, they have not yet been furnished.

Is it sufficient that the main artery of a limb should be preserved to support its life, and therefore the possibi-

lity of cure ? or, is the nerve passing to the uninjured
tissues equally necessary for the work of cicatrization?
What amount of soft parts can a limb lose without the
necessity of sloughing of those not detached ? That
question will, undoubtedly, receive a different reply
according to the limb injured, and the side on which
the tissues have been preserved. Can callus be reproduced when the broken ends are only kept in apposition by the soft parts, and without a continuity of
periosteum ? Can the muscles even serve as a mould
for the osteoplastic process, as M. Flourens thinks ?

Some precise results on these points would undoubtedly guide the surgeon in the indications which
he must often decide on very rapidly, and for which
purpose general rules would perhaps be more useful
than the various indications fixed upon by the
schools.

We shall confine ourselves to pointing out this
physiologico-surgical view which, sooner or later, will
not fail to bring about practical results.

6. CONSERVATIVE TREATMENT OF FRACTURES
OF THE THIGH.

As for conservative treatment, in 1848 it was that
of comminuted fractures of the femur in general, with
this difference, that the ball, occasioning a much greater
number of splinters and wounds of the soft parts than
in a simple fracture, they required much greater and
more elaborate attention. In but one hospital did I

see the *double* inclined plane made use of, and I must say that I was not altogether satisfied with its performance; adaptation appeared to me more difficult and less exact. On several occasions the deviation was so marked as to be recognisable at a distance. *The inclined plane*, by the very flexion of the thigh on the trunk, seemed to favour movements of the latter, and displacement of the upper fragment.

The *horizontal* position of the whole body, the head being almost on a level with the trunk, with or without extension of the fractured limb, shewed this inconvenience to a less degree; the extremity was sometimes laid on a long grooved pad and kept in the apparatus of Scultetus, and slight counter-extension obtained by a bandage fixed to the head of the bed. Turning of the foot outwards or inwards was prevented by another band round the instep fixed to the mattress on each side. In M. Roux's practice, the bandage of Scultetus was used without extension, the head kept very low. Blandin used the same bandage with extension and counter-extension. M. Monod, the inclined plane, etc.

The want of suitable adaptation of the fragments is the chief cause of the numerous failures in treatment of fracture of the thigh by the bullet. The *inclined plane* favours an outward curving of the limb, the horizontal disposes to sinking of the thigh, and in consequence to an antero-posterior curve. Compresses, suitable cushions introduced, changed as seldom as possible, obviate most conveniently this serious inconvenience; but the want of adaptation of the fractured bone is the rock on which even the most skilfully directed treatment ends by splitting. This continual presence of movable pieces of bone keeps up an inexhaustible depôt of suppuration, which, if it does not lead to death by purulent absorption, will cause it at a later period by exhaustion, and by a general restlessness which exaggerates the least pain, and renders all repose impossible.

The inclined plane has shewn much more than the horizontal position another very serious inconvenience, I allude to purulent deposits in the lower parts, and the necessity of making counter-openings at a distance from the seat of suppuration. Our notes are full of cases of counter-openings made near the hip-joint, and the escape of large quantities of pus.

I have seen employed, with equal success, *external incisions* near the wound. We have already discussed their importance in cases of tension (étranglement). Once four longitudinal incisions were made deeply into the right thigh, the swollen muscles protruded from the wound. After that the inflammation was subdued, and the prognosis was favourable. On another occasion, after a simple wound, the right was swollen to double its natural size, and livid. *Five incisions* were made like a crown round the limb, followed by protrusion of the muscles. On the morrow there was no appearance of gangrene; all went well. Some days later, more threatening of gangrene. M. Blandin speaks of a wounded man in 1830, on whom out of kindness he had abstained from enlarging the wound in the thigh; enormous swelling of the limb succeeded. Dampness and livid spots. He lost no time in making a long incision into the fascia lata, so as to relieve the inflamed parts from pressure. He was fortunate enough to see all the bad symptoms disappear.

For the reasons I have pointed out above, I ought to prefer the horizontal position in bed of the whole body, rather than the inclined plane. It is more easily applied, moreover, in time of war, inclined planes not being always at the disposal of the surgeon.

It may be objected, that the inclined plane exercises, by the weight of the leg, the surest and most continuous counter-extension ; *but is that counter-extension of great importance ?* is it always useful for the union of a fracture produced by a bullet, which almost always entails at the outset, from the removal of splinters, considerable loss of substance ? And is not the drawing up of the limb more favourable to the union than objectionable under these circumstances ? Dr. Stromeyer, whom I have already quoted several times, is, like ourselves, convinced of this, and he even attributes the reunions which he has seen in the Schleswig war, to having left the muscles to their own natural contractility.

That question has not been made sufficiently prominent here ; but it is important enough to merit all the attention of the surgeon.

One other precaution we cannot sufficiently recommend, consists in letting the broken limb lie on a cushion slightly raised to the level of the wound.

That pressure from *below upwards* has the advantage
of hindering the pus from working a way for itself into
the depending parts. The simple apparatus of Scultetus, with lateral pads, has not appeared to me sufficient.

It is to this cushion, placed evenly under the limb,
that I have attributed, in one remarkable case, the
good union of a tibia. Having been called on to extract two sequestra, comprising the whole circumference
of the bone for the length of 13 centimetres, I was so
fortunate as to see my patient perfectly cured without
deformity or shortening, after a daily dressing which
lasted for one *year*.

My attention was constantly directed to the lower
part of the leg. I heaped cushion on cushion below it
to combat the continual tendency of the limb to sink,
in consequence of the enormous loss of osseous substance, and a progressive emaciation. The complete
success which I obtained speaks in favour of, and the
importance of this precaution, which is as great for gun-
shot fractures as ordinary ones with loss of bone, and
the consequent flattening of the limb. In the case of
which I speak, the periosteum, largely opened in front,
was preserved at the bottom of the wound in the shape
of a gutter (demi-canal). A sequestrum had been
already cast off by suppuration before those I removed.
The case shews the extreme vigour of the bone-form-
ing power in a vigorous body of eighteen years of age.
The complete success, without shortening, was due in
part to the fibula, which, being intact, formed a natural

splint and a means of extension. This youth, cured now for several years, walks without lameness, and finds no inconvenience from going distances of twelve or fifteen miles.

The gutta percha concave splint was in general use in 1848. It rendered important services.

That complete envelopment of the limb resisted successfully the tendency to displacement, and diffused equally the pressure of the containing apparatus on its contents.

CHAPTER III.

WOUNDS OF THE KNEE.

INJURIES of this joint have been deemed by most surgeons as most serious, and as almost always indicating immediate amputation. They associate them, in this respect, with fractures of the thigh. For instance, in the Maison de Santé, in 1848, they had two cases under treatment, and both died.

1st Case.—A ball entered the left knee in front, and towards the inner side. It lodged in the internal condyle of the femur. Wound very small, one would have said it was caused by a buck-shot. On the 6th July serious inflammation followed, abscesses formed on all sides, and multiplied in spite of numerous incisions. They only extracted the ball on the 20th. Death from exhaustion on August 2.

The autopsy shewed no traces of phlebitis or metastatic abscesses.

2d Case.—A guard mobile wounded in a duel by a pistol bullet on the 27th July. Ball entered outside the right patella. They immediately enlarged the wound, and extracted the ball, situated behind the external margin of that bone.

Continuous irrigation. Same inflammatory symptoms as in the other case. Abscesses on all sides ; the irrigation has at least the advantage of allaying pain. Shiverings, delirium, and death, on August 20th.

In the face of the almost unanimous opinion of surgeons as to the extreme gravity of wounds of the knee, I venture to present some facts which seem in opposition to their experience.

Our observations on this subject are, however, too conclusive for us not to attach some value to them, even allowing for the fortuitous causes which meet sometimes in cases of the same nature.

Of 11 wounds of the knee which we have seen, in 4 the openings were exactly opposite each other, a line between them traversing the middle of the joint. Of these *three* recovered. In other cases, it seems that the ball has only injured the spongy ends of the tibia or the femur, without compromising the joint. Perhaps it is to that circumstance that patients owe the fortunate progress of their wounds. One can imagine, indeed, that a body possessed of so rapid a motion as the bullet could penetrate into the cancellated tissue without injuring much beyond the part immediately involved.

The harmlessness of some superficial wounds of the knee appeared to us sometimes due to the turning of the ball. Two cases of simple (sétons) wounds, one on the external aspect of the left knee, the other at the internal of the right, were followed by not the slightest inflammation of the joint and very slight suppuration.

The points of exit and entrance could not be united by a straight line without passing (the line) through the bone; it must be admitted, then, either that the ball traversed the condyle without producing suppuration, which is scarcely probable, or that it took a turn round the knee, between the bone and the skin. This is possible, when we consider that the ball is a contusing body, which can fold and pucker up the skin before piercing, so as to present, in consequence, two openings placed as we have described them. Penetration undoubtedly depends on the angle at which the ball impinges on the surface.

Twice we have seen cases where the ball passed laterally, so as to make an opening on each side of the patella. Both recovered. It is difficult to say whether the ball made a detour, or passed on the deep aspect of the bone. A remark of Dr. Stromeyer's speaks in favour of the frequent turning of the bullet in wounds of the knee. He mentions having frequently extracted balls from the popliteal space, which had entered by the side of the patella, and had passed under the skin to the extent of a third of the circumference of the limb.

The same author is struck by the slowness with which suppuration is established in a wounded knee, when the inflammation is energetically treated. Sooner or later, however, suppuration was set up by *foreign bodies*, and amputation was rendered inevitable.

An antero-posterior direction of the wound presents special dangers, from the tearing of the popliteal vessels. The following is a melancholy example : the bullet entered above, a little within the anterior tibial tuberosity, and went out at the popliteal space ; there was no apparent comminuted fracture. Slight hemorrhage was stopped by cold compresses. The patient appeared well and complained little. *July* 12*th*, Bleeding reappeared more violent than at first from the anterior orifice ; the patient looked pale and bloodless. A cold bladder was kept applied ; then ligature of the femoral artery about the upper third of the thigh. 17*th*, Gangrene beginning at the foot and spreading upwards. 20*th*, Great bullæ formed, and his condition seemed desperate.

In another case, to all appearance similar, there was a more happy result, explained probably by the German surgeon's remark, which we have just quoted. We have not seen a case of recovery in which the tibio-femoral joint was laid open so as to admit the external air. These are the cases which give to wounds of the knee their terrible reputation. There exists an intermediate region between the shaft and articular end of a long bone, which can be easily wounded, without, on the one side, splitting towards the former, and without, on the other, seriously reacting violently on the joint. That point is where, in early life, the diaphysis is joined to the epiphysis.

If, after these statements, the prognosis of wounds of the knee, taken generally, seems less invariably unfavourable on the other side, it must be borne in mind that the cure, when it occurs, seldom restores the knee to its original mobility, and that often there is nothing obtained but an anchylosis. The principal reason for determining conservative treatment here, then, is not so much the usefulness of the limb preserved as the danger of amputation in the thigh.

It appears to us that amputation should be had recourse to whenever the joint is so freely laid open as to admit the air; and that in less serious cases, especially if the ball has only injured the end of the bone, and if there are no decided signs of penetration or numerous splinters, they can attempt the preservation of the limb, deferring amputation till suppuration has become colliquative.

CHAPTER IV.

WOUNDS OF THE LEG AND FOOT.

THE best way of ascertaining the relative importance of these injuries is to compare one case with another, which have fallen under our observation, so I shall begin by giving an abstract of the cases which we met with in the hospitals.

The *first* was that of a man aged 25 years, with a wound in the middle of the tibia, which was broken with great mobility ; the muscles were much lacerated.

The dressing was applied in a simple concave gutta percha splint. Some days subsequently a large sequestrum was removed ; there was extreme mobility of one portion on the other, baring of the fragments, abundant but healthy discharge, the rapid granulation of the muscular edges causing protrusion. After the enormous loss of osseous substance, there could only be a few points of contact between the fractured portions. Those floating in pus are arrested in the process of ossification. Compresses and cotton wool filling up the interval between the dressing and the splint, poultices and a retaining bandage surrounding the whole. Five days later, the condition was very satisfactory ; the everted tissues had decreased ; the wound was contracted, and it was granulating from the bottom ; the discharge ceased to be fœtid. The patient's youth evidently greatly favoured a rapid cure. Three days later he was convalescent, the wound covered with healthy granulations.

The second case is that of an artilleryman with a fracture of the middle of his right leg. The ball entered behind and went out in front ; there being tension, vertical dilatation was practised, some splinters extracted, the limb was laid in a gutta percha concave splint and poulticed. Subsequently more splinters, one of which was as large as a thumb, were removed. Scultetus' bandage was applied, and an abscess was discovered on the internal aspect; this was opened, and gave vent to an abundant discharge.

Splinters continued to come away till September 25th, but his general health had been invariably excellent.

The wounds gradually closed, and by the end of November the patient was going about.

The third case, on June 25th, was a fracture of the middle of the right leg. The ball entered on the inner side, broke both bones, and went out behind. There was no bleeding, and the many-tailed bandage was applied. Subsequently, there was an unhealthy blackish discharge; gases collected in the seat of injury; the lower fragment was prominent, and threatened to pierce the skin. On July 8th, the leg was amputated. On examination of the upper fragments, they were found for the length of three inches, separated from each other like a **V**, and many smaller splinters.

9*th*, Strong rigor last evening and this morning; sutures removed ; skin and subjacent muscles gangrenous.   12*th*, No more shivering; gangrene confined to the flap.

13*th*, Rigors, sallowness, sweats ; given quinine. 16*th*, More rigors ; and in the evening death.

Fourth case; very like the preceding.   A man, æt. 60, had gangrene of the foot, occasioned, in addition to the influence of his advanced age, by the destruction of the arteries, the ball having pierced the middle of the interosseous space ; the limb was amputated.   There was immediate sloughing of the lips of the wound arrested by charcoal and powdered bark.   He was attacked by shivering.   Symptoms of pyæmia, and death.

Fifth case, æt. 35.   Fracture of the middle of the right tibia from the outside.   Escape of dark blood from the wound; mobility of the fragments ; slight denudation of the tibia; two counter-openings were made, and the many-tailed bandage applied.   The prognosis was guarded.   On the morrow he was going on well; the dressings were slightly tinged with blood. Six days later he was convalescent.

Sixth case, æt. 40.   Fracture of the right leg ; fragments very movable, protruding in the middle of a large open wound, from which flowed a thin and fœtid pus.   Pale complexion, sunken eyes, distressed aspect, sordes on the lips, and foul breath.   A double flap amputation was performed with a very bad prospect of success.

A comparison of these last cases teaches us that fractures of the legs by balls are more serious than ordinary ones, chiefly from the number of splinters, abundant discharge, and, consequently, risk of pyæmia, added to the long process of osseous repair, necessitated by the loss of substance, which leads in general to great mobility between the fragments, which is noted in the preceding remarks; and, moreover, bleeding is common enough. A ball which traverses the interosseous space readily implicates the arteries which, situated in that interval, are less able to escape than the vessels lying loose among the tissues.

It is hard to say, in a general way, what is the prognosis of a wound of the leg. However, it does not appear to us so unfavourable as to the German surgeon we have quoted, who considers that comminuted fractures of the leg, even when the fibula is unbroken, require immediate amputation. As our observations prove, even a double and complicated fracture of the leg is susceptible of cure. When speaking of the conservative treatment, we quoted the case of a youth whose tibia, removed for the length of 13 centimetres, was completely restored without any shortening.

From the lists of the Hotel des Invalides, it appears that from 1847 to 1853 there were admitted 158 fractures of both bones of the leg; 82 had been amputated, 76 cured without amputation; of the latter there were 22 in the middle of the leg, 20 above, and 34 below the middle.

From these lists it follows that fractures are more serious as they approach the knee-joint. Among those who had undergone amputation 31 had been operated on the same day, or the next. From the same reports it appears that a fracture of one bone has by no means the same danger as when both are broken. The prognosis of fractures, especially those of the leg, appears to us to depend directly upon the age and general health of the subject. There is really no comparison between the vigour of the bone-developing power at twenty in a healthy body, and that at fifty or sixty years of age.

The treatment will tend to preserve the limb in every case when the patient's strength will allow of the attempt, and when the destruction of bone has not been very extensive.

The mobility of the bones being in general very great, it will be necessary that the dressings should be changed as seldom as possible. It is in this case particularly necessary to support the limb beneath by a cushion, the height of which is proportioned to the displacement of the fragments and the thickness of the limb. M. Malgaigne, as is well known, obviates the anterior displacement by applying to the projecting bone an iron compress moving on a metal ring which surrounds the limb. The young man on whom we have seen this apparatus applied, did not suffer from it, and the pressure produced no untoward consequence. Some very painful startings, which deprived the patient of sleep, were probably unconnected with this mechanical pressure, and would not in any case have contra-indicated its employment.

External dilatations in the neighbourhood of the wound are especially appropriate in these cases on account of the unyielding nature of the tissues.

Poultices, especially when there is a tendency to constriction, appear to us only more useful in fractures of the thigh.

Wounds of the foot seemed to us more painful than others, probably on account of the inextenstibility of the tissues.   Here external incisions are less applicable on account of the closeness of the skin and bone.

Opiate poultices are in these cases the best calmatives.   The prognosis of wounds of the foot is generally favourable; even of complicated fractures of the os calcis, cuboid, metatarsal bones, etc., have recovered without subsequent stiffness.   In the campaign of Schleswig, amputation at the malleoli was twice performed after the method of Syme of Edinburgh ; one of these recovered, the other died from pyæmia.

I shall cite, in conclusion, as an example of a favourable case, the following—On June 23d, a ball entered the outer edge of the left foot, below the fifth metatarsal bone, it traversed the soft tissues of the sole, and went out between the first and second metatarsal bones, breaking the latter.   Some splinters were removed, with simple lint dressing the case went on well; on the 29th July, the patient had already been walking several days with his wounds healed, and no pain.

CHAPTER V.

Wounds of the Upper Extremities.

The gravity and progress of these wounds vary much, according to the parts injured; it is necessary to divide them into wounds of the shoulder, arm, elbow, forearm, hand.

One feels disposed, by a sort of general impression, independent of statistics, to consider wounds of the arm as less serious than those of the thigh. Our lists corroborate this opinion. Of 30 fractures of the arm and elbow, we count 17 cures ; whilst of 20 of the thigh, only 5 went on well.

Of these 30 cases, are—6 amputations,
3 disarticulations,
1 resection ;
that is to say, 10 serious operations, a proportion nearly double that of amputations at the thigh.

The difference, no doubt, is caused in a great measure by the greater facility with which we decide on the removal of the superior extremity.

Wounds of the shoulder are clearly the most serious. To the ordinary dangers of any joint wound, they add that of the neighbourhood of the thorax.

It would be hard to say, without extensive statistics, whether they are, on an average, less serious than those of the hip-joint. We have seen Dr. Legouest obtain cures of the latter, and preserve the limb. On the other hand, out of nine complicated wounds of the shoulder (including a disarticulation and an excision), we find four deaths. We have seen rapid cures when the ball had traversed the head of the humerus from side to side. This was the case in a vigorous man thirty-five years old ; the bullet traversed the neck of the right humerus. Little suffering, no alarming symptom, suppuration ensued, a splinter was extracted, the arm was kept immovably to the side, and in five days more he got up; no inflammation in the joint, no chance of splint. One would scarcely credit the fact that the humerus had been broken in its entire thickness.

2d case.—A boy of fourteen years of age, wounded in February. The ball traversed the left shoulder, near the joint. *July* 15*th*, He was well, but the shoulder and upper part of the arm had not its normal thickness. From both openings some pus flowed, and below is the cicatrix of a counter-opening. He cannot yet extend the arm, or put the hand to his head. He went out well, after having been so ill that they at one time proposed exarticulation of the arm.

3d case.—A wound of the left shoulder. In the region of the deltoid. The course of the ball comprises a line which traverses the head of the humerus.

Nevertheless, the progress of the patient was so

favourable, that one is obliged to suppose that the ball passed round that bone. There was abundant discharge ; a counter-opening made between the wounds and poulticed. A pad placed in the axilla, and the arm was bandaged to the chest. The best way to appreciate the relative importance and progress of these wounds of the shoulder will be to continue the enumeration of the most remarkable cases which have come before us, and to add occasional notes.

4th Case.—An enormous wound of the left shoulder from a shell ; almost complete denudation of the head of the humerus, which is merely attached to the trunk by the axilla and a little pedicle of flesh. General state very serious, all the symptoms of pyæmia, intellect embarrassed.

5th Case.—*June* 14*th*, Comminuted fracture of the humerus. The ball entered at the insertion of the deltoid, and went out near the axilla. The finger detected a mass of fragments as high as the head of the humerus ; the limb was placed in two gutta percha splints ; there was a copious discharge, but gases formed, and the pus became of a blackish hue.

7th July, The limb was amputated. On making the first flap, they found that the bone was splintered to its head, so they proceeded to disarticulate, which was done without any further complication. Union, by the first intention, was attempted by means of sutures ; cold compresses were applied. On the 10*th*, the sutures were removed, the wound only united above, but otherwise looked well. 11*th*, Slight shivering ;

was ordered sulphate of quinine. 16*th*, More rigors; the wound scarcely suppurating; dressed with ointments. 17*th*, Slight hemorrhage from the stump, subcrepitation at back of thorax on both sides, the wound looked ill, the discharge ichorous; on the 19*th* he died.

Autopsy.—On the left side, some adhesions. The lung exhibited on its surface a number of small abscesses. On the surface, next the diaphragm, was a group of a dozen abscesses, the size of peas, of a bright yellow, surrounded by a red circle; the lung itself, when cut into, appeared as in "hepatization."

On the right side, the pleura costalis attached to the apex of the lung was separated from the pulmonary lobe by a thin gelatiniform layer, in which blood vessels were distinct.

The lung, itself, at its base, shewed abscesses of a similar appearance to those on the right side, and was also hepatized.

The 6th case, somewhat similar to the preceding. On the 9*th June*, the ball entered in front, and went out behind the right shoulder, breaking the head of the humerus. *June* 30*th*, The latter was resected. They united the two openings by a transverse incision, by which the deltoid was raised in a flap. The head of the bone was smashed into four or five fragments easily removed, and the shaft was divided without difficulty.

The tying of the arteries occupied some time. Nothing untoward occurred until *July* 6*th*, when there were tympanitis and retention of urine. *8th*, Belly tympanic, but no rigors, and the wound looks well.   10*th*, Dyspnœa, pains in the left side of the chest, tubular respiration, and distinct bronchophony.   14*th*, Death. Autopsy, intense pleurisy with false membranes, and contractions.   Pneumonia in its second stage in the posterior lobe of the lung.

Fracture of the humerus at its middle is always a serious wound, and one which calls for amputation whenever the splinters are very numerous and the laceration of the skin extensive. English surgeons have ascertained in the Crimea, that out of 169 complicated fractures of the humerus, of 104 amputations 15 only died. The treatment is much more simple than in a fractured thigh. The arm, dressed with many-tailed bandages, is supported by a cushion placed between it and the thorax. A card-board concave splint is applied to the posterior aspect, and the whole is enveloped in a bandage which has a turn or two round the body.

What has been said with regard to fractures of the thigh is equally applicable here, taking, of course, into consideration the minor importance of wounds of the upper extremity, and the support for the fractured bones afforded by the thorax.

Fractures of the elbow can be cured with subsequent anchylosis of the joint. One example of this has been mentioned as a proof of the happy effects of continuous irrigation.

But we must not hesitate to perform ablation of the limb when there are great lacerations. As an example of the favourable result of amputation, when done in time, take the case of a boy, æt. 16, whose right elbow was entered by a ball. Considerable destruction of tissue; the fragments in numberless portions; no bleeding. Amputation of the arm at the place of election. Went on well for some days, when he had shivering, which continued. Subsequently, a sequestrum, representing the whole shaft of the bone, was removed, and the wound rapidly cicatrized.

Dr. Stromeyer, in 22 wounds of the elbow-joint submitted to his inspection, never once prescribed amputation. He substituted resection every time the elbow was the subject of comminuted fracture. When the ulna alone was involved, he only removed the head of the radius; when, on the contrary, the latter alone had been broken, at its upper end, he satisfied himself with extracting splinters, and smoothing off the end of the bone. With regard to the ulna, he merely removed the end of the olecranon, to prevent the awkward protrusion which that bone generally produces at a later period. When the articular end of the humerus was broken, he only excised what was absolutely necessary. That frugality in resection was crowned with the most brilliant success. Almost all recovered, with anchylosis of the elbow-joint and preservation of the movements of the hand. The limb remained immovable at an obtuse angle to the arm. The treatment generally continued for three months.

Wounds of the fore-arm and injuries to the hand in the campaign of Schleswig were very frequent; they generally got well without amputation. One, a fore-arm with both bones broken by a ball, which had pierced it from side to side.

The following are some cases observed in 1848—A ball pierced the middle of the left fore-arm, entering the palmar surface, and going out at the opposite side. Fracture of the radius and ulna ; many splinters were removed. The limb was placed on splints of gutta percha. Many more splinters were removed, but eventually it recovered.

Wounds of the fore-arm may be complicated with hemorrhage, especially if the bullet has traversed the interosseous space. What we have said of hemorrhage elsewhere is applicable here. We saw one case where the ball traversed the left fore-arm, breaking the ulna. Bleeding occurred thrice, and was stopped by compression. The patient recovered, with the movements of his arm scarcely affected.

Wounds of the hand.—In one case the ball left on the dorsal aspect a gutter-shaped wound. Continuous irrigation was employed, even throughout the suppurative period. Granulations appeared under the water and cicatrized. This case is mentioned to shew, that the continued application of water does not hinder the healing process. I shall also mention a wound of the left hand, which allowed of the palm, the index-finger, and thumb, being left intact.

In a third case, the ball removed the extensor tendons of the fingers. The flexors being unopposed, it was necessary to keep the part upon a flat splint.

The abundance of fibrous tissue about the wrist renders it more subject to sloughing and intense inflammatory reaction, therefore continuous irrigation is especially applicable.   For instance, on the 24*th June* 1848, a ball entered the edge of the second metacarpal bone of the left hand, breaking it and the third. From the first day continuous irrigation was applied, and there was neither swelling nor pain, but a healthy and moderate suppuration. *September* 17*th*, The patient was well, except that the third phalanx was close upon the middle finger, and there was some shortening of the second metacarpal bone.   The thumb was sound.

I add a last case, which shews that fractures of the wrist may be very serious, and lead to serious complications, in spite of remedies.   On *June* 24*th* a ball smashed the lower end of the third metacarpal bone, going out at the wrist-joint, and breaking the lower end of the radius.   Amputation was suggested, but declined. Continuous irrigation was employed.   This did well for some days, but abundant suppuration took place, phlegmonous erysipelas threatened to extend up the forearm, the skin was red, hot, tense, and painful.   Poultices were tried.   On *July* 8*th* the arm was removed at its middle, and the wound was united in part by first intention.

On *August* 6*th* the wound was cicatrized, but the patient was still in a bad way, weak, without appetite, sleepless.   He was sent away for change of air.

CHAPTER VI.

Wounds of the Pelvis and Genitals.

The prognosis of these injuries is determined differently by various authors.

In the German campaign of 1849, these wounds, every time that they involved one of the pelvic organs, were attended by most serious consequences. In June 1848, on the contrary, they were often cured. In most of the recent cases we met with, the ball entered at the fold of the groin, and escaped at the hip.

In none was there laceration of the arteries or sciatic nerve. For an explanation of this favourable result, we must suppose that the ball turned on the internal surface of the ilium, and injured no important organ. For, how else could these recoveries be explained? A ball entered in the left groin, went out at the hip, probably by the sciatic notch. The symptoms were trifling; a slight pain on moving the hip-joint, scanty discharge, and the peritoneum was evidently untouched. On 15th the patient was up, and could, without pain, support himself on the limb of the wounded side. 21st, Is in a fair way towards recovery;

however, he suffers slight pain, probably from some splinter. The leg of that side was slightly shortened by muscular contraction. Second case almost identical. Cured. Third also similar. Fourth, The ball entered above the right groin, went out at the margin of the hip, probably turned by the neck of the femur. The bleeding was stopped by compression, vigorous suppuration followed, and a rapid recovery. In one case the ball, which had penetrated by the sacrum, went out at the pubes, and caused infiltration of urine. Osseous fragments penetrated the bladder, and were removed by the lithotrite.

M. Jobert, among twelve wounds of the pelvis, counts eight almost instant deaths; three recovered, and in them the ilia were pierced. Most recorded gun-shot wounds have been in the right iliac fossa. In two patients who recovered, there was no suppuration in the course of the ball; in a third there was with exfoliation of a small piece of the ilium.

All the pelvic wounds, treated in the Val de Grace in 1848, appear to have recovered.

We seldom meet with injuries of the genitals. One case displays a remarkable peculiarity. The ball went in above the glans, traversed the corpus cavernosum, thence went round the pubic arch, and went out at the right hip.

The penis and scrotum were swollen, congested, and there was bleeding from a scrotal artery. About fifteen days after there was no tenderness to the touch; on the upper part of the prepuce was an opening the size of a bullet.   It was impossible to determine the direction of the ball in the penis and deeper tissues.

Some days later, it was necessary to give exit to pus by a counter opening near the testicle. The scrotum was swollen as in double orchitis, with the skin and penis red and glistening, with frequent erections. The wound of the prepuce looked well, he urinated easily, and a scanty discharge came from the posterior opening. On the morrow he was better, the discharge was bloody, the erections still occurred ; he was ordered camphor and opium. Six days later, another incision was made on the right side. It healed up, with perfect recovery. In another case, the ball carried off the lower part of the glans.

CHAPTER VII.

Wounds of the Head, Face, and Neck.

Wounds of the head by fire-arms, in which the projectile produces general cerebral disturbance and many splinters, are different from those produced by direct blows. Seldom healing by the first intention, their cure is often preceded by exfoliation of bone, and suppuration on the surface of the brain.

Frequently the inflammatory process slowly spreads from circumference to centre; however, the cerebral symptoms may appear very suddenly. Take the following, for instance:—A case of wound of the cranium was going on well. Within twenty-four hours there came on complete coma, stertorous respiration, widely dilated pupils, the patient could not be roused by the strongest stimuli. The trepan was applied to the right parietal bone, and matter escaped, containing cerebral substance—he died. The slowness in the occurrence of inflammatory symptoms is owing, no doubt, to the density of the cranial covering, which is only involved by slow degrees.

The following is an example where the motion of

the nerves was more involved than the functions of the brain : A superficial wound of the left parietal bone, followed by delirium for four days, paralysis of the right arm, but not of the leg. A few days after this, movement returned, but the hand still could not close tightly on any object.

The treatment of wounds of the head consists in combating inflammation.

The internal remedy in 1848 was tartar emetic in cautious doses, largely diluted. Trepaning in two cases did no good. This operation, so popular in former days, is rarely performed now, except to remove splinters, and to give exit to any fluid to which the symptoms of compression may be referred. The following case is an instance.

A deep fissure of the right frontal bone.—At two small points one could see the dura-mater and the pulsations of the brain ; and it was evident that a layer of pus covered the dura-mater. The patient wandered from time to time, otherwise the symptoms were not very serious. He was trephined, and died ; on examination, pus was found spread over the brain.

Wounds of the face are most serious when they involve the frontal region and the eyes. A ball may pass right through the face, and the wound heal, without great exfoliation. A ball has been known to pass from the middle of the face to the mastoid region, and only occasion facial paralysis.

At the capture of the Mamelon Vert, an officer of the Turcos had his face completely taken off by a ball

—chin, mouth, nose, cheeks, eyes, nasal fossæ, tongue, all had disappeared—only the skull and neck remained. The wretched man breathed, and, by his changes of posture, and the significant movements of his hands, shewed he was aware of his condition. He sank twenty hours after.

Dupuytren considered wounds of the lower jaw very serious, but subsequent experience shews that such injuries are often cured. In one of our cases, where the bullet passed through the lower jaw, from right to left, the wound, supported in a rigid bandage, did well, and the patient could speak as well as ever. In another case, the bullet carried away four teeth, and was flattened out on its passage. In a third, the ball went in at the mouth, which was open, broke some teeth, passed under the tongue, and went out under the maxilla.

Dr. Stromeyer thinks that to facilitate the reproduction of bone in the lower jaw, one should only remove adherent splinters when they have been detached by suppuration, so that the periosteum may be preserved as much as possible.

CHAPTER VIII.

PENETRATING WOUNDS OF THE CHEST.

THE first symptoms which occur after penetration of the thorax by a bullet are often very alarming.

The wounded man undergoes great nervous shock, breathes with difficulty, coughs, expectorates more or less blood, pure or frothy.

Our remarks as to shock making death appear imminent are especially applicable to this class of wounds. In twenty-two cases, hæmoptysis was present, but it is not a pathognomonic symptom of penetration; it may be only a casual complication of some superficial injury.

In the chest, we have often to refer the favourable progress of the case to the circuitous course of the bullet, when the wounds of entrance and exit are in a line which traverses the lung. The following case is the only one where this was proved by post-mortem examination :—

A young soldier was received, on June 17th, into the Hospital of the Faculty for a wound of the chest. The ball entered about the middle of the clavicle, breaking that bone, and passing out behind through the infra-spinous fossa of the scapula. No accident to the side of the chest, no hemorrhage, no paralysis of the left arm. He went on well for some days; then, about the eighth day, he had feverish rigors, and died at five in the afternoon. The autopsy shewed the injuries already described, and that the ball had *not pierced the thoracic cavity ;* it had traversed the axilla in its long diameter, penetrated the serratus magnus and subscapular muscles, then broke the scapula ;. it had passed just below the vessels and nerves, which lay in some broken down tissue. In another less conclusive case, the ball entered at the margin of the left breast, followed the curve of one rib, and went out below the axilla without injuring the thorax. It was only necessary. to make one incision in the middle of its track to allow of exit to the pus. Dr. Hennen followed, by anatomical examination, a bullet which passed into the thoracic cavity, going out at a point just opposite that of its entrance, and leading one to believe that the man's body was pierced from side to side, the more so as he spat clots of blood.

The prognosis of these injuries is difficult to determine in any certain manner, as no one sign enables us to recognize exactly the nature of the internal injury.

Balls passing through the thorax have not always produced death. On the other hand, we have seen a wound entirely superficial complicated by fatal pneumonia. The lists of the imperial Maison de Santé give four deaths out of nine penetrating wounds ; our proportion would be (dead and doubtful) about twenty-two, that is to say, two-thirds.

Taking into consideration the consecutive complications, the prognosis of wounds of the chest does not seem less favourable comparatively than those of the abdominal and cranial cavities, and even fractures of the larger limbs. John Hunter has shewn that wounds of the chest produced by a bullet are generally less dangerous than those by a cutting instrument. That difference is sufficiently explained by the bruising nature of gun-shot wounds in general.

Three principal points present themselves in the treatment of wounds of the chest : 1. *Primary hemorrhage ;* 2. *Effusion;* 3. *Consecutive phthisis.*

The first of these is due to the rupture of some important vessel, an intercostal or pulmonary artery, and must be distinguished from bleeding resulting from injury of the pulmonary tissue, which but rarely by itself induces fatal consequences, while the former may do so very speedily. Phthisis is only developed at some future period, but pleuritic effusion is a most frequent secondary cause of death. The autopsies of which we have notes have generally shewn pleurisy with adhesion.

The fluid found then, or that which flowed on tapping, was very abundant, somewhat like thin milk, generally mingled with flakes of pus, sloughs, and clots, etc. Sometimes the coagulable part of the liquid was adherent to the pleura, in the form of a gelatinous layer. Evidently a simple increase of the secretion caused by inflammation, mingled with all the secretions and debris of the wound, opening into the cavity of the pleura.

The course of these wounds presents three different stages, corresponding to the three principal causes of death. If the patient has overcome the first shock, if hemorrhage has been arrested, he may enter apparently on convalescence, and, if not warned by experience, one would say he was almost cured. It is often at this stage of delusive prosperity that I have seen appear, without any accidental cause, the gradual symptoms of effusion, a return or paroxysm of the fever, burning heat of skin, thirst, increasing dyspnœa, no respiratory murmur, dulness. Tapping is more directly indicated in traumatic effusion than when the latter is the result of idiopathic inflammation.

There is in the first case a centre of suppuration, which causes a sufficiently free outlet by the respiratory passages ; thus the discharge has been abundant, and followed by rapid relief, but unfortunately this occurrence is rarely curative. In almost all the cases where Larrey performed paracentesis, it only produced a passing amelioration, and ended in death. Dupuytren considers that he hardly cured by it more than three or four in fifty. The lung, according to him, requires weeks and months to regain its volume.

Larrey enlarges on the subject of empyema ; he gives, among others, interesting details of one case where effusion was diagnosed with accuracy without the aid of auscultation, and confirmed by the autopsy.

The subject is so important that I shall shortly re-
capitulate the symptoms by which he ascertained the
seat of the effusion :—

1. Evident palpitations on the right side.

2. Right ribs much less inclined and more horizontal, the intercostal spaces larger.

3. Œdematous swelling, which Larrey thinks a
characteristic symptom.

4. The patient could not lie on his right side without suffocation.

5. Abdominal organs driven downwards, and prominent.

6. No pulsation on the right radial, and hardly
any in the axillary, carotid, or crural arteries.

7. Left hypochondrium motionless.

8. Countenance slightly swollen on the left side.

Bloody expectoration lasts several days, or even
weeks ; when the inflammatory reaction is established,
inspiration produces a pain, more or less acute, at a
point corresponding to the wound.   The symptom
which in favourable cases lasts the longest, is an obsti-
nate, dry, and often very painful cough.

The treatment is more medical than surgical. Save
excision of a ball lying on the surface, removal of a
piece of rib, paracentesis, and the occasional ligature of
an artery, the treatment comes under the care of the
physician.   In 1848, it consisted of repeated bleedings
(seven and ten times in Jobert's practice).   If we declared ourselves against bleeding in ordinary wounds,
and after a fatiguing campaign, we acknowledge its

utility when the wound has its seat in the central organ of the circulation.   Dr. Macleod, founding his views on Crimean experience, recommends it in these cases, otherwise it was rarely employed during the war.

The patient drank solutions of nitre, with or without tartar emetic, in limited doses.

The diet ought to be less substantial than in any other wounds.   Some hygienic irregularities by the wounded in 1848 appeared to me the cause of very serious pleuritic relapses.

There is a question relative to the treatment mentioned here, that is, the necessity of keeping, from the very first, the external apertures of the thoracic wound hermetically closed.   Larrey, Dupuytren, and almost all surgeons after them, agree in recommending perfect closure, especially with the view of arresting hemorrhage.

In the cases which came under my own notice, the apertures, having had time to take on inflammatory action, were left alone.

The circumstances which occur in cases of penetration are as follow : — The lung sinks immediately, more or less, according as the projectile has destroyed large or small air tubes.   That sinking produces immediately in the pleural cavity a vacuum, which is filled by the external air.   At every dilatation of the thorax by the pectoral muscles and the diaphragm, that air-filled space increases, to decrease again at each expiration.   The air inspired by the mouth can penetrate directly by the wound into the pleural cavity, and only imperfectly distends the parenchyma of the lung. The dilatation of the thorax will produce in the wound a kind of suction, which will favour the flow of blood, and the latter easily escaping, the bleeding is kept up. The hermetic closing of the wound, in hindering the external flow, will also dry up the source of the hemorrhage. After some days, the tissue of the lung will be swollen, and on its way to cicatrization ; the air henceforward no longer penetrating from within outwards, the inspirations restore the volume of the lung, and by so much diminish the pleural cavity. And if empyema should demand paracentesis, the puncture of the latter will be less troublesome than the original wound would be, had it been left open, as the trocar will be entered at a point not corresponding with the lung wound, which will have had time to cicatrize. On the 24th June, a ball entered about the middle and outer side of the xiphoid cartilage, traversed horizontally, so as to go out under the axilla, fractured the seventh and eighth ribs in two places at its entry and exit, so that the fragments possessed great mobility, and were seen to move at each respiration ; there was effusion of the same side. *July 12th,* He died. Autopsy shewed fracture of the seventh and eighth ribs in three different places, with considerable effusion of a reddish fluid on the left side.

On the 24th of June a ball entered at the external edge of the left shoulder-blade. Not extracted. Dyspnœa, intense orthopnœa, which were somewhat relieved ; pericarditis came on ; bleeding and blistering were used. On July 10th, delirium ; on 12th, he died. The autopsy shewed baring of the sixth rib, with fracture of its lower edge ; the ball lodged in a mass of pulmonary debris very near the rib, flattened and covered with a piece of stuff like cloth. The lung healthy, save an intense pleurisy ; pleura covered with false membranes ; effusion between the lung and pericardium ; nothing on the latter, or the heart.

Case 3.—A man æt. 17, the ball entered by the external edge of the right pectoralis magnus muscle, near the axilla, did not pass out ; there was pain on pressing the right scapula, and the finger recognized a slight non-fluctuating prominence, which made one believe in the presence of the ball. Respiration well performed ; no fever. Except slight cough, and spitting of a viscid mucus, he might have been called convalescent on the 9th July, when there was slight fever and excitement following an imprudence in diet. 11*th*, The fever gone, respiration tolerably free, always slight cough, with expectoration of mucus, streaked with blood. 17*th*, Fever, which seemed symptomatic of a commencing effusion, was discovered, well marked on the right side.

A puncture by a trocar was made between the eighth and ninth ribs, in the middle third of the intercostal space, which gave exit to eight spittoonfuls of thin fœtid pus, or, rather, purulent water the consistence of thin milk, and the odour of rotten eggs. The evacuation, accomplished by percussion of the chest, became more resonant. The exudation was not reproduced, but the patient was, nevertheless, but slightly relieved. The ball was subsequently removed at the prominent and painful point on the shoulder. He recovered.

Case 4.—A young man had his left index finger smashed by a ball, and had it amputated. Another ball passed through his chest from right to left. Immediately there was intense shock, nervous tremor, and hæmoptysis. About ten days after this, he breathed without pain, the cough, which continued, caused a painful resonance in his chest ; he was convalescent on the twentieth day, after having some symptoms of empyema.

Case 5.—A round wound as if *punched* out over the right shoulder blade ; the ball remained in. The general shock at first, and other symptoms, as abundant hæmoptysis, dyspnœa, etc., were so alarming, that the doctors thought him dying, but notwithstanding all this, he recovered, though with painful cough, and the physical signs of effusion.

Case 6.—A young man, whose upper angle of the left scapula was pierced by a ball, which traversed the muscles of the back to escape above the middle of the right clavicle. There was at first hæmoptysis, but he was going about in fifteen days.

CHAPTER IX.

WOUNDS OF THE ABDOMEN.

THERE are not so many recorded as we would expect, considering the large surface the abdomen offers to the ball. M. Baudens, among 164 wounded, had but 7 of the belly. M. Jobert records 11 cases, of which 8 were almost immediately fatal. But wounds of the abdomen are not always necessarily fatal. Indeed, of 7 cases, of which 5 were clearly penetrating, 3 offered good hopes of recovery; there was no vomiting, the stools were uninterrupted; the only sign of deep-seated abdominal disturbance was the jaundiced hue of the body, which, by its somewhat deeper tint, could be distinguished from the sallowness of pyæmia.

One of my colleagues saw two cases of penetrating wound of the liver. In one, the ball remained there; in the other, there flowed from a double aperture a laudanum-coloured fluid, evidently a mixture of bile and blood. The following cases got well, contrary to the opinion of Dupuytren, who considers penetrating wounds of the liver absolutely fatal. The ball entered in front, under the right false ribs, and went out in the loins. The patient progressed favourably for the first few days. No vomiting, regular stools, slight jaundice, then clammy mouth, no rigors or fever; favourable prognosis. In another, the ball entered at the left side, was extracted under the skin of the back. He went out cured.

In M. Baudens' practice, there were 7 wounds and 4 recoveries. 1, Garde Mobile, wound in left side, *cured*. 2, Right hypochondrium wounded, fracture of false rib, omental hernia, ball lost, *died*. 3, Right hypochondrium at back part, ball lost in the belly, hepatic fistula, *satisfactory*. 4, Wound under the umbilicus, on its way towards cure was lost by imprudence in diet. 5, Right epigastric region, with hepatic fistula, *convalescent*. 6, Penetrating wound, *died*. 7, Wound of omentum, *cured*.

The English statistics relative to the wounded in the Crimea shew 100 deaths out of 127 penetrating wounds; that is to say, but a sixth were cured.

That much greater average mortality is undoubtedly explained by their including all those who survived some hours, whilst in Paris they often passed several hours in the ambulances before being brought to the hospital.

The comparative danger of visceral lesions depends on the organ wounded. The unexpected recoveries after some wounds scarcely allow one to doubt that a ball may pass round the abdomen without touching the intestine. Another circumstance which must influence the prognosis is, whether the intestinal canal was full or empty.

The treatment of these wounds should have for its chief object—keeping the injured intestines completely at rest, so as to favour, if possible, the co-aptation of the intestinal membranes with each other, or, at least, those of the visceral wound with the abdominal parietes, and the formation of an artificial anus.   This end, it is true, is but seldom obtained.

Larrey relates a remarkable case of this in detail. The ilium had been cut by the ball.   The two ends of the gut had separated, protruded, and were puffed up, the passage of matter was interrupted.   Larrey passed a loop of thread into the mesentery, and by it kept the wounds of the bowel against that of the skin.   The two ends of the intestine adhered, and there was complete cure.   In the Crimea antiphlogistic treatment was abandoned, and for the most part replaced by *opium*.   The English surgeons are especially loud in the praises of this drug for obtaining immobility of the bowels, associating with it a severe diet and injections.

This treatment, followed in the Crimea, is that which the English prescribe regularly in their own country for all inflammations of the abdomen.   They realise, at the same time, the extreme importance of only nourishing the wounded for several weeks on liquid and soft food, the absorption of which should only necessitate a slight peristaltic action.   Perhaps we may favour this immobility of the intestines by a bandage lightly passed round the body with an opening corresponding to the wound.   This will have, besides its mechanical effect, the advantage of rendering the abdominal parietes less sensible to changes of temperature, the action of which on the peristaltic movements is well known.

We have cited above, a case where splinters of bone were removed from the bladder by the lithotrite.   In other cases, the ball drew with it into the bladder various hard bodies, as pieces of money, buttons, etc. When small, they were removed through the urethra by progressive dilatation.

THE TRANSPORT OF THE WOUNDED.

*New method of Transport—Plaster Dressing—
Starch Dressing.*

The military surgeon is not usually obliged to undertake the removal of the wounded with his own hands. His duty, if he follow a company into action, is to direct this transport, and to do so, he must be acquainted with all its details.   He should therefore exercise himself in the various modes of transport, as in such cases, merely looking on never affords such intimate acquaintance as individual experience.   It is especially necessary that he should be able distinctly to indicate to the bearers in what positions to hold their bodies, arms, hands, etc.   It is scarcely needful to remind surgeons that a broken limb, for instance, will present a totally different prognosis, according to the more or less suitable manner in which the conveyance has been effected.

The manner of transport will, of course, vary according to the seat of the wound, its severity, and the distance to be passed over. If the injury is in the head, the chest, or one of the upper limbs, and if it does not appear very serious, the patient will be able to go a short distance on foot. In the two first instances (the head and chest), he should be supported by two hospital assistants, one on each side. His arms being uninjured, he will be able to place one round the neck of each bearer. Each of the latter should have the arm across the back of the patient, and the hand placed under the opposite armpit; care must be taken to lean the hand more against the arm than the chest when the wound is thoracic. If it is an arm that has been wounded, it may be lightly supported on the chest by tying a handkerchief round it. The bearer placed on the side of the sound arm will support it under the armpit, whilst the patient holds him round the neck; the other, placed on the wounded side, will encircle with his arm the body of the patient, taking care that there is no pressure on the wounded arm. In general, walking ought not to be attempted except *for short distances, and in cases of slight wounds.*

Whenever the wound is serious, the patient ought to be carried, whatever be the seat of the injury. The human body is too heavy, and of too unwieldy a shape, to be borne along for any length of time by one. If there are *two bearers,* the removal is much better ensured, and less exposed to accident. I mean by this that, when it can be avoided, a wounded man ought not to be entrusted to only one supporter. With regard to fractures of the limbs, especially of the lower ones, no removal must be attempted until some means have been applied for holding the jagged ends firmly in position. This precaution is of the greatest importance. It must be remembered that it is always better that the splint should be too long than too short, and that, if possible, it should include the neighbouring joints. Finally, the wounded limb should be fastened to the sound one, the latter being the natural splint of the former. If one of the lower limbs has been fractured, it will be better to do no more than place the patient where he may be sheltered from projectiles, and to delay his removal until the necessary means have been provided for properly setting the wounded limb, rather than attempting to remove him before applying this dressing. A litter is always preferable to carrying in the arms ; it ensures the greatest stillness to the body, and the movements of the bearers cause only very slight reverberation.

The most commodious litter consists of two wooden poles, from 10 to 12 feet long, with strong canvas sacking firmly stretched between. The poles are held from $2\frac{1}{2}$ to 3 feet apart at each end, by two others placed transversely. Extempore litters are founded on the same principle ; the poles in that case are two muskets, or better still, two strong branches from some neighbouring tree. The canvas is represented by the leather straps of the shoulder belts, by cords, a soldier's greatcoat, or by shirts found in the haversacks. In the latter case, the poles are inserted into the shirt, and a number are placed one above the other. There may be some doubt about the solidity of a litter of this kind, and before placing the patient upon it, it should be tried by a soldier to ensure all being in proper order, for nothing can be more distressing than to see a litter break down, or bend, under the weight of a wounded and suffering man. A good sized ladder also serves very well as a litter. There is one point on which I must strongly insist, because it is one too often disregarded, viz., never to proceed to the removal or transport of the wounded before having agreed with the bearers on their line of conduct. This is the time for the surgeon to prove that he is a calm determined man, who silences all useless talk and discussion —cautious and reflecting in the midst of the excitement and thoughtlessness by which he is often surrounded. I cannot sufficiently impress on the military surgeon all the dignity, all the greatness of his mission, on an occasion like this; but that mission to be well fulfilled, requires a certain amount of head work, and if the surgeon is not always exhausted or wearied by great physical exertion, he assuredly is by the labour it has been to him to maintain his coolness, to think and foresee before others act, and to display a resolute but calm will in the midst of agitation.

I shall not allude to the use of cacolets, and ambulance cars. Their use belongs more to general regulations than to surgery, properly so called. I shall merely remark that the cacolet *with rests* is very rarely of use even in wounds of the trunk, of an arm, or almost of the foot ; but that it is impracticable in wounds of the abdomen, in fractures of the thigh, wounds of the knee, and compound fractures of the leg. In cases such as these, the cacolet *bed* has rendered excellent service. In order to prevent the shaking as much as possible, it is desirable to make the patient sit, or lie upon very pliant cushions, and for this purpose I can recommend those made of India rubber, which I have frequently made use of.

First Dressing of Fractures of the Limbs.

According to the careful statistics, and the discussions we have gone through in this work, the field of battle is destined to see the number of immediate amputations increase rather than diminish. The consequences of such a surgical system are very apparent in thus increasing the number of mutilations ; it adds to the number of useless or nearly useless members of society, and involves expense for the necessary support of this army of invalids. Thus a question, at first purely surgical, touches on questions of more general public interest. In the face of this serious question, whether, in the first place, to preserve some limbs, but to lose many lives ; or, in the second, to save a greater number of lives, but to make many invalids ; the surgeon naturally asks, if art cannot afford some means of lessening the one danger without increasing the other. Following the progress of surgery in gunshot wounds, the point to which the surgeon ought to turn his serious attention is *the application* of the first dressing to the fractured limb, intended to keep the ends from rubbing during removal. We shall describe in this chapter two modes of dressing—1st, The plaster of Paris and dextrine; and, 2dly, The application of a special dressing of our own invention. The first, as is well known, consists of strips of linen, dipped in a solution of plaster, bound rapidly round the limb; it forms in drying a sort of unyielding shell, in which the limb is firmly imbedded. Here are some of the principal rules pointing out, as briefly as possible, the manner of using the dressing. The plaster ought to be pulverized, well calcined, and of a middling whiteness, rather than brilliant white ; it should be mixed, adding water until it is the consistence of cream. In this state it solidifies in from five to seven minutes. It is better at first to put too much water than too much plaster. Through this mixture are passed strips of linen of from once and a half to twice the circumference of the limb ; it is then wrapped round the limb, and the layers repeated two or three times or more, but not oftener than six so as to avoid too great a weight. The bandages are to be applied directly on the skin, which has first been smeared over with oil.

A surgeon, with the aid of three experienced hospital assistants, can apply such a dressing as this in ten minutes. It ought to have openings through which the surgeon can observe the progress of the wound and apply the requisite remedies. These openings are usually made in the following manner. A string dipped in oil is placed round the wound in a circle, and the plastered bandages placed carefully over this. When the removal has been effected, and it is desirable to open the dressing, this can be done by cutting the bandages within the circle of the cords with a pair of strong scissors, made for the purpose, in the shape of a gardener's shears. Care must be taken to include the joints immediately below and above. The limb should be placed in an intermediate state, neither too much bent nor quite stretched out—the latter is much the most convenient for removal, but the patient cannot always bear it. The plaster dressing, although intended for fractures of the limbs, is also applicable to wounds in other parts of the body—such as fractures of the jaw, wounds of the pelvis, etc.

The advantages of these dressings are—in the first place, cheapness ; 2d, the facility with which plaster can be procured everywhere ; 3d, the rapidity with which it can be applied ; 4th, firmness, and the safe-guard it affords against jolts and shaking during re-moval ; 5th, porosity, which serves to absorb liquids. On the other hand, the inconveniences of this dressing are—first, its weight; and in the second place, the difficulty, often very great, of opening it without shaking the limb, which is always prejudicial to reunion.

The movable-immovable apparatus of Seutin is on the same principle as the plaster dressing. It consists, as is well known, of a starch bandage applied to the limb, then cut in two shells, which having been placed on the body, take its form, and can be re-applied at will. This dressing may also have openings which, however, are not so necessary as the former method—the dressing being divided into two movable portions. The apparatus of Seutin is lighter than the plaster, and the scissors cut it more easily; but, on the other hand, it is not so thick, and offers less resistance to outward shocks. Besides this, it is much easier to find gypsum than starch at one's disposal in the field, and the latter takes longer time to harden. The apparatus of Seutin is, therefore, better adapted for stationary treatments, while the plaster dressing does best for cases of removal. It must be confessed, however, that the starch dressing possesses some great advantages in the treatment of comminuted fractures by gun-shot wounds. The union of these depending, in great measure, on the immobility of the limb, it is obvious that the advantages must be great of an apparatus, moulded upon it, fitting perfectly, and offering sufficient resistance to prevent any flexion of the limb. Impressed with the danger to reunion of transport under disadvantageous circumstances, I invented and had made, that it might be tried in the war in Italy, an apparatus for fractures, which I shall now proceed to describe, and submit to the opinion of surgeons.

After having been tried in the army of Italy, this apparatus was introduced into the military hospitals of Paris and Turin, and into the Spanish army. It consists of a number of splints and vulcanised India

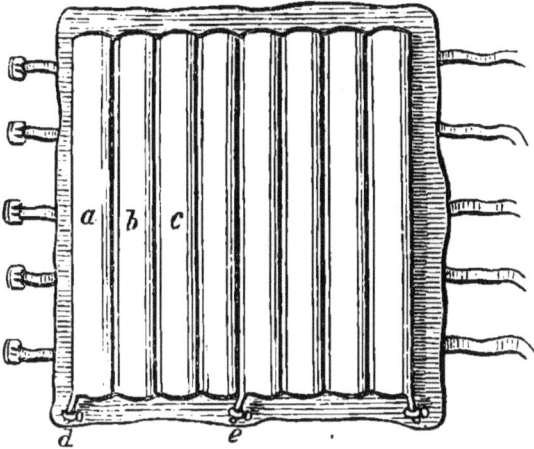

Fig. 1.

rubber air cushions (figure 1, *a*, *b*, *c*). The cushions are joined together, and form one whole. At the extreme end of each cushion, or pair of cushions, there is a plug for inflating it (*d*, *e*.) The splints, to the

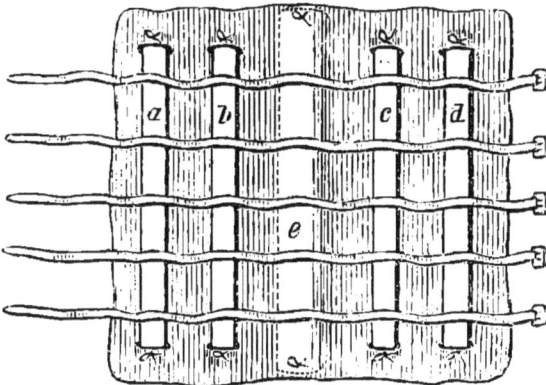

Fig. 2.

number of five, are bound in strong canvas, which

o

serves as a covering to the cushions (fig. 2). I preferred five splints to three, the usual number, that it may entirely surround the limb, and thus ensure more entire immobility during the removal. The apparatus varies in the size, and number of cushions. In the simplest, there are 4 cushions, about 50 centimetres in length to 15 in breadth, inflated by 2 plugs and 5 splints of the smallest size, with 3 straps.

This very simple form is intended *especially* for wounds of the knee, the leg, and the foot. The other more complicated apparatus which I am about to describe, was the object of my most careful study. It is intended for fractures of the thigh, and for complicated wounds of the knee. Here the cushions must be 6 or 8 in number, longer and narrower than the former, 70 centimetres in length to 6 in breadth ; they may be inflated by either of these plugs. There are 5 straps. But what particularly distinguishes this apparatus from

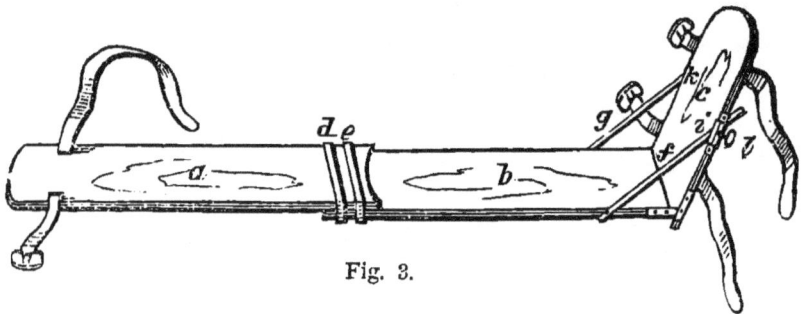

Fig. 3.

the other is the long splint (fig. 3). It is composed of several pieces, intended to admit of its extension, and yet form as firm a support as if they were all one piece of wood. The splint is composed of two half splints and of a sole for the support of the foot (a, b, c), the two halves joined together by two smooth brass rings, can slide over each other without losing their respective directions. Upon the back of one of the boards is a series of notches (encoches à cremaillère)

Fig. 4.

(fig. 4, a, b, c), into which fastens a spring or hook, fixed at the end of the other board. A very simple mechanism thus affords a firm and gradual lengthening, both being effected by simply drawing out the splint. The footpiece is fastened to the lower board by means of a lateral hinge, which, being firmer than an ordinary one, allows at the same time of the sole being

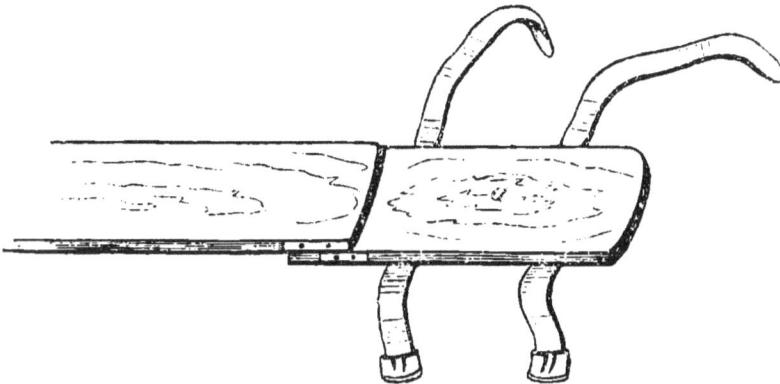

Fig. 5

brought down entirely (fig. 5, a). Two brass supports, fixed laterally (fig. 3, g, h), can then be detached, go into two openings formed by a metal plate, and slide into it without any difficulty (*i*, *k*); they are stopped by a screw. The sole is, besides, furnished with canvas straps, intended to keep the foot firm, and which are buckled to the lower surface of the boards, so as to avoid all painful pressure on the foot. A longer stirrup, placed at the upper end of the splint, fastens it to the body. When this apparatus has been used, it is easy to shorten the long splint by folding

Fig. 6.

it up to the exact size of the other simple splints same in the thickness, which must necessarily be double (fig. 6.)

Manner of Applying this Apparatus (Dressing).

The following is the manner of its application on a healthy body supposed to be wounded in one of the thighs :—

1. The person is laid in the horizontal position. 2. The dressing is quickly opened, and spread out smoothly by the side of the thigh supposed to be wounded. 3. The long splint is drawn out in both directions to its utmost length without withdrawing it from the canvas sheath in which it is enclosed. 4. The wooden sole is lowered horizontally. 5. The

whole apparatus is passed under the injured limb, so
that the sole may be properly fitted to the foot, which
is firmly fastened by means of the leather strap.
6. The upper end of the splint is fastened to the pelvis
by the long strap. 7. After the long splint has been
thus firmly applied to the limb, the assistant must
stoop down and inflate the cushions successively by
the plugs. This requires strength, and some degree of
perseverance. 8. The whole apparatus is then drawn
completely round the limb, taking care that the cushions
meet in front, and all firmly fastened together by five
straps ; in doing so, making use of the support of the
left knee. This is the time when the buckles are put
to the test. 9. Finally, the sound limb is firmly
fastened to the wounded one. Experience has curiously
enough proved how entirely the leg is secured in this
apparatus from any shocks from without. One may
even sit down upon it violently, when fixed and buckled
up, without the leg experiencing anything beyond a
slight increase of pressure. All shocks from without
spread immediately, by the law of undulations, through
all the cushions, and produce only a trifling direct effect.

Subsequent experience will decide what amount of
general use this apparatus merits. It does not pretend
to take the place of the starch and plaster dressing,
which are much less expensive ; but it is superior to
these from its much greater suppleness, and can also
be used as long as the cushions are not injured or
torn.

DISINFECTANTS.

THE importance of taking every precaution against
the decomposition of sloughs and discharges cannot be
possibly overrated.

The products of the decomposition of such organic
matter are dangerous and hurtful in two ways ; they
are irritating and they are poisonous.   It is needless to
enter into the chemical nature of these products of decomposition to furnish evidence of their mephitic character.   The absolute necessity of cleanliness in respect
of such matters in the mass, is fully recognised, but it
appears to me that sufficient attention is not paid to
the dealings with such sources of infection in detail.   It
is futile to attempt to "keep a ward sweet," if the
sources of contamination be not individually disinfected.
Nothing in my visit to the Paris hospitals after the
battles of June 1848, struck me more than the contrast,
both as regards general appearance and local condition,
between those patients whose wounds were systemati-
cally disinfected by the liberal application of nitrate of
silver, and those who were treated by poultices and
other dressings, which retained the discharges in contact
with the inflamed and superheated part ; and I have
to acknowledge my indebtedness to M. Jobert de
Lamballe for a most useful practical lesson in the disinfection of wounds.   Instead of quoting cases in support of my proposition, namely, that disinfection of

discharges is of first importance, I would rather say, let any one treat two cases of whitlow, for example, as follows : one in the ordinary method, by unlimited poultices; the other, by minutely and carefully removing every particle of exfoliating epidermis—a substance, by the way, especially prone to putrefaction — and of slough, and by syringing the sinuses with a proper disinfectant (of course, supposing that in both cases the requisite incisions have been equally made), and then let him compare the time occupied in the cure of each.   That the more rapid cure will be by the latter method of treatment, I will venture to predict.

It may be well to state that considerable misconception appears to exist in reference to the capabilities of disinfectants, so called, and that blunders in their application are consequently committed.   To sprinkle the floor of a ward, in order to purify its atmosphere, with a solution of chloride of zinc, is wholly inefficacious. To apply a solution of permanganate of potash, by means of lint wetted with it, is to deprive it of the power of acting.   It is evident that offensive and poisonous vapours diffused through the air of a chamber can only be attacked by a volatile or æriform agent such as chlorine.   It is equally clear that a disinfectant which acts, by itself undergoing decomposition on contact with organic substances, at once loses its virtue when soaked into linen cloths, or mixed with poultices.

Disinfectants may be arranged in two classes—1st, Those which act by preventing decomposition, or by modifying it, by chemical union with the substances liable to decompose; 2d, Those which absorb or neutralize the volatile or gaseous products of decomposition.

To the first class belong—

Chloride of zinc,
Nitrate of silver,　　　　⎫
Permanganate of potash,　⎬ in solution.
　　　　　　　　　　　　　⎭
Nitric acid.

To the second class—

Hypochlorite of soda or lime.
Chlorine.
Charcoal.
Iodine.

Chloride of zinc may be placed at the head of disinfectants of the 1st class. The solution should contain from one to five grains in the ounce of distilled water. It may be syringed over the sores or into sinuses, or applied by saturating the immediate dressings. ·

Chloride of zinc sometimes causes a good deal of pain, and brings out small pin-head pustules, but I have never known real mischief result from its use. It can be employed with signal advantage as a gargle in salivation, and as an injection into sinuses in the maxillary bone, from which the discharge is invariably most offensive, as well as distressing to the patient.

The solution of permanganate of potash may be used of a strength according to circumstances. It can only be applied with the syringe, since contact with cloths, sponges, and the like, decomposes it. Permanganate of potash is said to act by yielding oxygen to the organic substances with which it comes in contact. It is, as might therefore be imagined, in some measure *corrosive*, and if used too strong, causes the tender granulations to bleed.

The solution of nitrate of silver must also be applied with the syringe. It differs from the permanganate in having astringent and styptic properties. It should be used not weaker than ten grains to the ounce of water, but twenty grains or more is frequently not too strong.

Strong nitric acid is absolutely invaluable. By powerfully coagulating the tissues, it is capable of converting the nidus and pasture ground, so to speak, of phagedæna into an unchangeable and an insensible shield to the next threatened parts. It arrests the phagedænic process in the same way that congelation of the fluid in the cells of a galvanic battery would arrest galvanic action. A respite is given which affords the opportunity for the recovery of the ground that has slipped from underneath the patient's feet, and of strength which may enable him to combat and restrain, within due limits, inflammatory changes.

Hypochlorite of lime and of soda (chlorinated soda and lime) yield chlorine when exposed to the air, and are thus the most convenient atmospheric disinfectants, or, disinfectants of the second class.

It is true that weak solutions of these salts may be used as lotions, but they have the disadvantage of being less chemically stable than chloride of zinc ; and thus it is liable to happen that the nurses and attendants may continue to use what is really an inert wash in place of a disinfecting solution.

Freshly burned charcoal, a material always obtainable, placed in wire gauze cages, or in muslin bags, over the sores, or in the bed, is capable of absorbing, and, it is supposed, of inducing, the oxidation or destruction of certain volatile matters. It may be used where the vapour of chlorine would give rise to undesirable irritation of the lungs or larynx.

Iodine acts in some measure as chlorine. It may be employed by enclosing a small quantity (Ʒij) in a chip box, which can be suspended over the patient's bed. My colleague, Mr. Campbell de Morgan, by whom disinfection by iodine vapour is used in the Middlesex Hospital, tells me, that the method is the invention of Mr. Hoffman of Margate.

SURGICAL APPLIANCES.

BANDAGES.

THE simplest form of bandage consists of a strip of cloth varying in length and breadth with the purpose to which it is to be applied, prepared for use by being rolled up tightly from one end; hence its name of ROLLER.

Writers on the subject term the mass so prepared, the *head*, and its free end, the initial portion. When rolled from one end it is a single, when from both ends a " double-headed " roller.

ROLLERS for the limbs are usually single-headed, and vary in breadth with the circumference of the limb. A lean leg requires a narrower bandage than a stout one, but the average breadth may be said to be from $2\frac{1}{2}$ to $3\frac{1}{2}$ inches. Eight yards of such will be required to bandage a leg from the toes to the knee.

To bandage a leg.—Set your patient with his leg in the horizontal position, and place a book or other convenient support under the heel. Lay the back of the initial portion in front of the ankle joint, turn the bandage round the latter, and bring it down in front of the foot, then fold back the loose end so that the next turn will cover and so dispose of it tidily. Now turn round the foot close to the roots of the toes, and continue doing so till you cover the dorsum. Unless you are about to apply a starch bandage or other apparatus likely to hurt the skin, you need not cover the heel but continue up the leg. The swell of the calf will prevent the turns lying smoothly unless you *reverse*, which is done by stretching the bandage at an acute angle (not a right one as in circular bandaging) from the limb, and then gently drawing it back towards you, at the same time turning your knuckles upwards, this will fold the bandage on itself. When the round part of the leg below the knee is reached, a few circular turns may be made, and the bandage pinned or tied.

These reverses not only make the roller lie smoothly on the limb, but when made opposite each other, present a neat and artistic appearance to the eye.

For the arm.—It is sometimes necessary to begin at the fingers, with narrow strips suited to their size, or they may be included in one roller. Or the bandage may be turned round the wrist, then through the cleft between the thumb and fore-finger over the hand, folding back the initial portion neatly, and beginning to reverse immediately above the wrist-joint.

Great care must be taken that the pressure is equable, especially that it is not greater anywhere than at the lower end of the limb, as under such circumstances it is very apt to interrupt the venous return, and do irreparable mischief.

Bandaging with a roller should be practised till all sizes and shapes of limbs can be covered quickly and smoothly without any irregular pressure or tendency to loosen. The roller is among bandages what the scalpel is among knives, it will do anything a skilful hand requires of it, and is capable of many useful modifications.

On trunk and groin.—Place the back of a roller six yards long on the lower part of the abdomen, pass it twice round the body, then round the thigh as high as possible, then again round the body, repeat this alternation till the bandage is exhausted ; this is a Spica bandage.

This application of the roller is much used for retaining dressings or compresses on the groin or inguinal region, after operations for hernia.

An analagous bandage may be applied round the thorax and one shoulder, or round the neck and one shoulder, as when we wish to keep dressings on the latter without much circular compression.

For the head.—They are generally double-headed, as it is convenient to use both hands, and by an intertwining of the bandage to obtain a degree of fixedness which a single-headed roller could scarcely achieve. Their most common use is to retain dressings or compresses. Place the portion between the heads of the roller on the dressings or vertex, the first turn may be round the cranium or under the jaw, as most convenient, the heads will again and again change hands till they are exhausted, or a sufficient application made. As the direction of the turns must greatly depend upon the position of the object for which they are applied, it is needless to lay down any precise rules, but it is well to

Fig. 7.

remember that many turns under the jaw will prevent its movements, and I have often seen a man's head so elaborately bandaged, that his jaws were fixed, and his eyes blindfolded (fig. 8).

As for the proper application of a roller, it is necessary to pass the hands freely round the parts bandaged, and as in some circumstances it is inconvenient to do so, the roller may be cut into pieces of lengths equal to what a turn and a half round the part would have required.   These pieces or bandelettes are then stitched by their middles across one long strip, overlapping each other one-third or one-half of their breadths, this MANY-TAILED BANDAGE is laid flat on a pillow or bed, the limb gently raised and placed upon it in such a way that the long perpendicular strip may be exactly in the middle of its posterior aspect, then the ends of a bande-lette are taken one in each hand, crossed over the limb, and drawn with sufficient tight-ness to prevent their

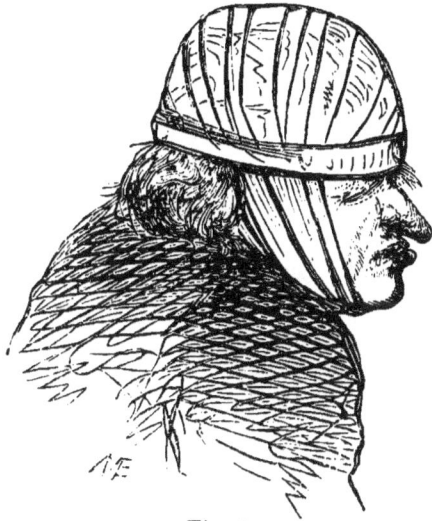

Fig. S.

shifting. Should there be a wound or any prospect of the bandage being soiled by discharges, the bande-lettes at that part may be left unstitched, so that a fresh one being fixed to each, when the dirty are pulled out, the clean are drawn in. This many-tailed must not be confounded with the four or six-tailed bandage, which consists merely of a piece of cloth, of a suitable breadth, torn from each end to within a short distance of its middle. The tails of this bandage may be rolled when applied, and the untorn portion laid over the dressings as in the double-headed roller.

The T bandage consists of a strip of cloth, with one or two bandelettes stitched to its middle, and is used to keep dressings on the perineum.

Another bandage in which the bandelettes are

longitudinal, is the *retaining* bandage, which is useful as
a means of drawing together the edges of wounds, and
has been used instead of pins in operations for remedy-
ing "hare lip." Three or more narrow strips of bandage or tape are attached at equal distances to the end
of a roller, and a corresponding number of slits cut in
the latter in such a part of its length, that, when the
bandalettes are passed through them, and the bandage
applied, they will be over the wound.

When a wound is transverse, as in the thigh, a
little management with either one or two rollers will
enable the surgeon to have both bandelettes and slits
at right angles to the turns of his bandage.

The best material for these simple bandages is common calico. The unbleached variety does well, but
those who are particular about appearances are annoyed
by the profusion of its threads. Those who habitually
wear bandages on the leg, and who can afford it, generally use the stocking-net, which is soft, elastic, lies
well on the limb, and cannot do much harm, however
slovenly it be applied. It costs about sixpence a yard.
A good medium between the expensive kind and the
common cotton is a loose soft fabric sold by most
instrument makers at the moderate charge of sixpence
for as many yards.

Many who suffer from varicose veins and ulcers of
the leg, who can neither afford elastic stockings, frequent bandaging, nor the rest which would alone enable
their legs to do without efficient support, find a good
substitute in calico, made adhesive by some simple

resinous material, applied in strips like a many-tailed bandage. The calico should be unglazed; the strips an inch, or an inch and a quarter, in breadth, and long enough to reach quite round the limb and overlap. For covering a leg from the toes to the knee, a piece of strapping one yard long by half a yard broad will be required. The strips should be an inch or more in breadth, but those for the calf must be narrow, in proportion to the swell of this part.

The surgeon then takes a strip by each end, draws the *unplaistered* side across a heated Italian iron, or a pitcher full of hot water (fire is apt to *roast* the plaister, and destroy its adhesive qualities), then applies the middle of the strip to the back of the ankle, crosses and smooths the ends over the instep, a second is now placed under the heel, and smoothed stirrupwise up each side of the ankle; having thus insured a proper support for the cutaneous vessels which are generally, in cases such as we suppose are under treatment, very full about the malleoli, and in front of the joint, he begins to apply his strips from the toes to the knee, crossing them as he did the bandelettes of the many-tailed bandage, and leaving as neat a series of digitations as in a well put on roller.

As the strapping will be kept for some days on the limb, it is as well to put a bandage over it, or a long stocking; and should there be an ulcer, it is well to cut an opening in the strapping over it, to allow the escape of any discharges.

Over it may then be placed a piece of lint, with the

ointment or other dressing selected by the surgeon, kept on by another strap, which not being overlapped by its neighbours, may be removed and reapplied from time to time.

Adhesive plaisters have the great advantage of not shifting or getting lax, and are therefore very useful as *retaining* bandages (see treatment of wounds) for dressings or splints, especially in young children whose lithe and restless bodies wriggle away from almost any other fastenings. But their unyielding nature requires that straps, when used as bandages, should be applied with the greatest caution ; and fortunately, such is their cohesive power, that, when merely applied in the long axis of the limb, they will adhere with sufficient firmness to answer the surgeon's purpose.

For extension they should be thus applied :—The plaister being on coarse unglazed calico, cut strips of such a length, that, when they are applied longitudinally on the limb, there may be about nine or ten inches overplus, and of from $2\frac{1}{2}$ to 3 inches in breadth. Apply one strap on each side of the limb, or more, as the exigencies of the case seem to demand; and should they cross tender points or projections of bone, protect the latter by stuffing in a little cotton wool. Next apply a roller over them. Should the limb be thin and lean, and there seem danger of their shifting, get a little common starch or batter, or the white of a raw egg, and smear the bandage with it. Now take the spare ends of the plaister straps, twist each into a rope, and fasten them to your splint, taking care that nothing is compressed between them. When they have been on a short time, you may pull upon them, with the comforting assurance that while you have all necessary control over the limb, your extending media neither circularly compress nor excoriate.*

All who have treated fractures of the thigh, know how important the last-mentioned points are. Before I learnt to extend by this method, I was once desired to attend a man who had been three months confined to his bed by a fracture of the middle third of his thigh bone. The day I saw him he had tried to walk with crutches for the first time ; they slipped, the bone was again broken, and another weary time of probation was before the poor fellow, with this aggravation, that in all the situations where bandages are commonly applied under such circumstances, the former ones had left deep ulcers, and it cost me no little pains and anxiety to keep his limb in position. Now, longitudinal straps of adhesive plaister would have saved him pain, and both of us trouble and anxiety. Young children are often the subjects of fractures of the femur; they require frequent change of posture, and are scarcely ever at rest. For them plaister is certainly more suit-

* I believe the word extension, in the surgery of fractures especially, is subject to much misapplication, and inappropriate in properly treated cases. Because, after the coaptation of a broken bone, there is really no more extension required unless there be again displacement. The perineal band on a thigh splint, for instance, no more extends than the odontoid ligaments do, but, like them, it is a check to displacement.

able, for the reasons already stated. I was once, when a student, desired to remove a long splint from the thigh of a child eighteen months old; it had been applied apparently very well three weeks before; but the practitioner, in his fear of displacement, bound it so tightly that one turn of the bandage had cut down to the bone, the flesh had closed over it, and much trouble was required, not only in its removal, but in the subsequent treatment of the limb.

RIGID BANDAGES.—It is sometimes desirable to render the bandages rigid; and we can do so by saturating or covering them with some material which hardens on drying.

Many substances have been lauded from time to time. Larrey recommended a paste made of flour and the white of eggs. Professor Hamilton of America uses wheat-flour paste. Alfred Smee introduced a mixture of gum-arabic and whiting. Baron Seutin of Brussels, generally thought the highest living authority on the subject, has used starch in the treatment of fractured bones since 1834. Mr. Erichsen uses it largely in the University College Hospital. Others, again, prefer British gum or dextrine. F. d'Arat gives the following formula for preparing it for bandages :—Dextrine, 100 parts ; camphorated brandy, 60 ; water, 50.

To bandage a leg with dextrine about 1 lb. is required.—Provide two roller bandages, each eight yards in length. Put one to soak in a basin of water while you unroll the other and saturate it with the dextrine mixed with warm water to the consistency of treacle.

Roll this bandage again, taking care as you do so to exclude any lumps which may have formed. Now, having placed something to protect the bed or carpet, and something to elevate the heel, apply your wet bandage to the foot in the usual manner, but covering the heel; some put on this first bandage dry, but it shapes itself to the heel and ankle better when moist. Should there be any tender points, cover them with a little cotton wool. Having finished with this bandage, take the one which has been soaked in dextrine and apply it in the same manner—at every few turns smearing some more of the treacly fluid over it. Now, keep the bed-clothes off the limb with a cradle, and let it lie quietly till the dextrine dries, which takes several hours ; it is a good plan to put it on towards the afternoon, and it will be firm and hard before the patient wishes to move in the morning.

Plaster of Paris may be applied in two ways. The skin having been protected as described —

1st. Take an ordinary roller, or what is better, one of that loose fabric alluded to at p. 208, unroll it, rub the plaster of Paris into its meshes, and then roll it up again. Then soak it in water for a little, and apply it in the usual manner ; or put it on dry, but sprinkle water over each turn you make. The advantages of plaster of Paris over dextrine or starch are, the ease of its application, and the rapidity with which it hardens. The disadvantages are chiefly those of bulk, of liability to be spoilt in the mixing, and hardening before being properly applied.

2d. Having applied your bandage, or bandages, and having everything likely to be wanted within easy reach, mix the plaster of Paris in a basin gradually, with about an equal weight of water, or add water till it becomes of the consistence of cream. Then smear it over the limb with a broad brush or the hands, bearing in mind that it hardens very rapidly, and that one supply must be made at a time ; after it has begun to solidify any additions of water entirely spoil it.

If put on too thick it cracks and falls off in flakes.

DRESSING OF WOUNDS.

Although the modern treatment of wounds consists rather in protecting them from any interference while the processes of repair are going on, yet much may be done by judicious attention to expedite and perfect their cure. The reaction which followed the meddle some surgery of the last century and first quarter of this, has been rather excessive, and the affected contempt for dressings, the throwing of all responsibility upon "Nature," has been carried rather too far in the last ten years. Mr. Jones of Jersey told me that he attributed the remarkable success attending his operative practice, to his habit of seeing his patients (not *dressing* them), every four hours, for several days and nights after serious operations.

Meddlesome surgery is childish, but slovenly surgery, which allows dressings to become dry and harden over a wound, the secretions to accumulate and emit a sickly odour, is a great deal worse. The simplicity of modern dressings happily renders both unnecessary and unlikely to occur.

As this is not a work on systematic surgery, it would be out of place to enter into a disquisition on the healing of wounds, but we may take a bird's-eye view of them, thus :—The accidentally inflicted wound a surgeon is called to, will, in all probability, partake of one or more of these qualities—

Wounds.
{
Incised . . . may heal by first intention.

Lacerated, contused { may do so in part, though in proportion to the laceration or contusion there will be sloughing.

Punctured { whether caused by the rapid course of a ball, or the rapid passage of a sharp, hard instrument, may heal by first intention, but is not very likely to do so in its whole extent.
}

They may be *complicated*—

If incised . . by divided vessels or tendons.

If lacerated, contused . { by destroyed tissues, which will slough and act as poisonous foreign bodies.

If punctured { by foreign bodies forced in, and destroyed pieces of tissue.

But, as already stated, the modern treatment of wounds is expectant, the injured tissues get the benefit of the doubt, and, however we may have to change our treatment in the subsequent progress of a case, in the first place we act as though expecting immediate union, and therefore, we endeavour to remove the *obstacles* of union, the chief of which are—

General.		Local.	
	Close atmosphere.		Foreign bodies.
	A naturally unhealthy constitution.		Blood.
	Any extremes or sudden changes of temperature.		Admission of air.
			Motion.
	Disordered digestive functions.		Debility.
	Debility.		Heat.

We therefore cleanse out the dust or dirt with a stream of water or soft sponges; if deep and narrow we explore with probes or fingers, lest any extraneous substance be present. We close it to exclude the air; we lay the injured part on a splint if it is likely to be moved; we endeavour, by stimulants at first, and then by nutritious food, to remove the weakness caused by the injury, and we keep the patient cool and in the best air obtainable.

Wounds of the Skin generally reunite by the first intention, but if the integument has, either by accident or design (as in Taliacotian operations), been raised in a flap, great care must be taken lest part or the whole of it should slough, and while the necessary cleansing of the raw surface from clots, etc., is performed, and the parts adjusted, the margins should not be pinched with forceps or fingers. Dieffenbach attached great importance to the venous return of blood from such a flap, and endeavoured, by dividing any arterial twigs entering it, to prevent any considerable quantity of blood being thrown into the flap, " which would cause it to perish from repletion. When such an accumulation takes place, we must encourage a gentle flow from the lower edges of the flap."* Should the deep surface seem inclined to ooze blood, a compress may be laid over it ; but great pressure or undue stretching may induce congestion or ulceration, either in the whole length or part of a flap, and it is often more prudent not to attempt attaching the skin margins to each other, than endanger them by forlorn-hope attempts of immediate union. Much may be done towards their approximation in the after treatment of a case.

Mechanical aids to Coaptation.—The edges of skin wounds may be kept together by strips of adhesive plaister or by stitches. The latter are generally introduced by means of needles, of which the average length is about two inches and a quarter, flat in the back, and tapering to a long sharp point. The back at its middle should be a third broader than the eye end, so that the wound made by it may be large enough to transmit, without tugging, any thread which the eye will admit. Of late, some modifications of the eye end have been made to accommodate metallic threads. After many trials of each, I have returned to the old surgical needle, with a broad eye, as the best and most generally useful. In flat surfaces, or those where the skin is difficult to get hold of, a greater curve should be given the needle, so that it may hook up the tissue it has to transfix.

Some, for greater freedom of manipulation, prefer a needle fixed in a permanent or temporary handle, as

---

* This may be one reason why the most carefully formed flaps from the heel sometimes slough, though generally accounted for by the arteries being cut too high up.

the Porte Aiguille. The latter enables one to displace the needle after it has perforated the tissues, and saves the trouble of unthreading, which under certain circumstances is very troublesome. All these objects are combined in Professor Simpson's tubular needle, which he uses for sewing up vesico-vaginal fistula. This needle is first passed through the margins of the wound which are to be brought together, a wire is then pushed along till it appears at the further end of the tunnel, it is then held with a forceps till the needle is withdrawn by the route it had already traversed. Serre-fines, or little spring clasps, are not much used in this country.

Threads may be now-a-days divided into metallic and non-metallic. Of the latter, silk is in most general use, waxed. For fine wounds French twist is excellent. Where there may be a complication of threads, as in the operation for cleft palate, it is convenient to have them of different colours, to prevent confusion when tying them.

Fig. 9.

Of Metallic Threads.—The best is No. 28 iron wire ; it is cheaper and stronger than silver, which is more

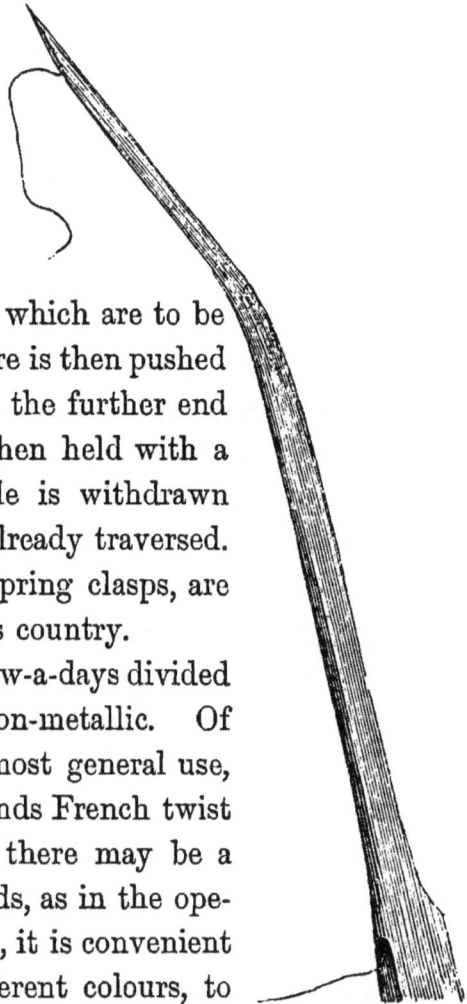

apt to get into kinks and break. Nearly every wire has been tried, but in Germany and this country surgeons are quietly settling down into using iron. The stitches for closing ordinary wounds are generally inserted at short intervals or interruptions from each other, hence the term interrupted suture. They should be sufficiently numerous to keep the edges well together without crowding; should be sufficiently deep and far from the margins to secure a good hold; and, above all, should not be too tight. When one is seen to mark the skin, it should be slackened, otherwise it will cause ulceration.

Some surgeons introduce more sutures than others, some pass them only through the skin, while others transfix the deeper tissues, which is essential in some operations, as in those for ruptured perineum.

After each stitch is inserted, it may be drawn together and tied; an assistant should gently hold the lips of the wound together, so that the stitches may be opposite each other—any fat that protrudes may be pushed back or clipped off with scissors; and, before closing the last stitch, any blood which may have collected in the wound should be gently pressed out. Some attach so much importance to oozing of blood as an obstacle to union by first intention, that they leave the sutures uncut and untied, with the raw surfaces exposed to the air, or with merely a piece of wet lint laid lightly over or between them; when, after the lapse of a few hours, they become glazed with a layer of lymph and white blood globules, they are considered to be in a more favourable condition for union than when the wound was first inflicted. Although in theory this practice is often mentioned, most surgeons are reluctant to put a patient to the additional pain of closing the wound, when he has flattered himself all operative measures were over. Some express direct disapproval of it, but it is well, without dogmatizing on what must always be a matter of opinion, to reserve it for cases which shew an unusual tendency to ooze. Should the stitches not keep the margins of the wound quite in apposition, or should the parts appear to need some further support, strips of adhesive plaister may be laid transversely in the intervals between the sutures.

Plaisters for keeping wounds together should be as little irritating in their nature as possible. The strongest, which is spread with resin, unfortunately is so. The soap plaister is but of little use if much traction is required, and the transparent isinglass tissue is somewhat expensive; moreover, the discharges are apt to loosen its hold of the skin. If the plaister seems to interfere with the cicatrization of a skin wound, nothing is easier than to protect the latter with a narrow strip of lint spread with spermaceti or lard, what Cutler calls a Protective Bandelette.*

Fig. 10.

The *Sutura-Sicca* of the ancients, rarely used now, may be applied in various ways—in straps, as already described, or the plaister being applied on each side of

* Cutler's Dressing and Bandaging, p. 11.

the wound, the sutures may be passed through it instead of the skin. Kid leather spread with adhesive material, and perforated with "eyelet" holes along its margin, may be laid down in the same manner, and the wound laced up. I have used this in long wounds on irritable subjects, where I was desirous to avoid sutures, and been well pleased with it, but it is difficult to adapt it properly to the surfaces, and it scarcely, in ordinary cases, repays one for the additional trouble. This was tried by Roux, but is spoken contemptuously of by Malgaigne in his "Médecine Opératoire."

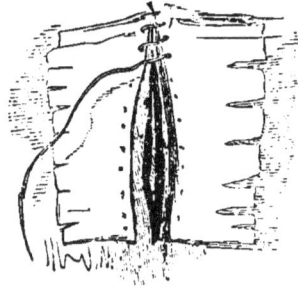

Fig. 11.

In some cases, even of amputation, we may substitute straps for stitches. Lint spread at each margin for about an inch with plaister, and cut into strips, combines the useful traction of the strapping with the painlessness of the water dressing. I have classed straps with the interrupted sutures, as practically it is impossible to sunder them. Most surgical writers advise their being substituted for stitches in closing wounds of the scalp, as the irritation of the latter has sometimes been the supposed cause of erysipelas. But unless a pretty wide space is cleared of hair round the wound, the straps will not adhere sufficiently to draw the tight skin sufficiently close. Few persons, especially those most liable to cuts on the head, like their mishap to be observed, and it has always seemed to me that the speedy union generally obtained by sutures was less likely to be followed by disagreeable symptoms, than a wound less efficiently closed by straps, which are necessarily short and liable to slip.

Although these remarks apply especially to skin wounds, they are more or less adapted to any tissue, and since the metallic suture has come into general use, the old rule of merely stitching skin surfaces together, seems less adhered to every day. The form of suture requires some modification, however, for particular circumstances and situations—thus, in very mobile parts, such as the lips, where a number of small muscles converge into one common terminus (the orbicularis oris), we must obtain rest for the part while it is undergoing repair. With this view, therefore, we endeavour to combine with the suture a splint, or mechanical rest for the part, by means of long rigid pins inserted at some distance from the margin, and traversing the cut edges nearer their deep than superficial aspect. If a thread, which should be soft thick silk, be

twisted round these pins, it will be seen to pull on the tissues external to the entrance of the pin, and any muscular drag must also cease at this point, leaving the wound in the calm centre of a circle.

Fig. 12.

Twisted Suture, as used for hare-lips or face wounds, is thus applied:—Take a pin and enter it at such a distance from the edge of the wound that it may, by a gradual slope, pierce the latter near its posterior or deep aspect; pass it straight through the other edge, and let it reappear through the skin at a similar distance to its entrance on the other side. Care must be taken—1st, that the edges do not pucker ; 2d, that the pins are sufficiently deep ; and, in operations on the lips, that the red margins are in a line. If the wound has been made for the removal of an ulcer

Fig. 13.

or tumour of the lip, the subsequent appearance of the part will, of course, depend upon the amount of tissue removed, but if for the removal of some deformity, then it is of great importance that no notch should be left to lessen the perfection of the cure. It will be frequently necessary to combine the twisted and the interrupted suture, and also to make

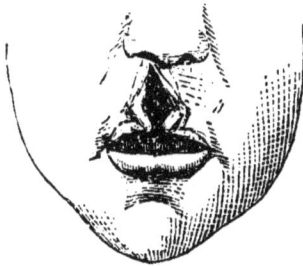

Fig. 14.

some allowance for the drawing up of the margin. A very little calculation and knowledge of tissues suffices ; thus, instead of entirely removing the parings, one may let them hang till the wound is adjusted, and then trim what is wanted of them into shape, or the knife may be carried thus, or thus (figs. 14 and 15).

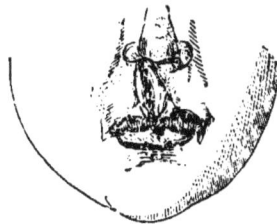

Fig. 15.

These little tags are very delicate, and must not be pressed on too much or they will slough, so it is better to take a fine needle and put a stitch in.

How long ought you to leave in the pins ? is a

question often asked, but difficult to answer, as the amount of irritation caused by them varies with the patient. The best rule is to watch carefully after the first forty-eight hours, lest the excitement which accompanies the healing process should pass into inflammation and suppuration. If they are covered entirely with thread, it is impossible to do this, but the evening of the third, or morning of the fourth day, will be soon enough in most cases.

In removing the pins, first gently rotate, so as to loosen them, and draw them gently out, taking care not to jerk them or to leave the part unsupported. If very curious to see how the wound is progressing, you may now remove the threads, but it is better to have patience, to smear them with collodion, and let them drop off like a scab in a day or two. The wound is often so firm that no further support is required; at other times it is only joined in parts, and looks inclined to reopen entirely—whether it does so or not, of course depends upon whether the action is a healthy one or not, but, in any case, it ought to be closed with strips of plaster, or a pin should be passed through again, and if there seems a want of activity, a few drops of red lotion * be dropped upon it, or a little charpie tucked in under the strap. Those who have not had opportunities of observing how comparatively large gaps fill. up under these circumstances, may occasionally despair

---

* Red lotion is composed of two grains of sulphate of zinc to every ounce of water, and red lavender sufficient to give it colour and fragrance.

of their operation being successful when they see the margins quite asunder in parts, but I have seen a lip merely adhering by a small part of the prolabium, the rest remaining open, throw out granulations which fused together, leaving little or no cicatrix.

Some obtain the requisite rest for the wound by buttons, as in Mr. Wood's plan. A good substitute for the common straight pin is the nursery or safety pin, which requires no clipping, may be obtained electro-plated at a very cheap rate in any hardware shop, and by its double bar prevents the thread from pressing upon the wound. It will be found that in many cases nursery pins of the smallest size, which only cost $4\frac{1}{2}$d. a dozen electro-plated, make an excellent substitute for the twisted suture. In parts likely to swell, as the scrotum, it will be found that a fine thread of caoutchouc yields to the swelling, and tightens again when it subsides, so as to hold the edges together without cutting.

Fig. 16.

The continued Suture, or stitching the lips of a wound together as women do a seam, has been recommended for wounds of the intestine, and in long wounds of the skin it saves time and trouble. Where the stitches are laid very close together it is termed the the glover's suture. These methods are useful where very close apposition seems necessary to prevent the entrance of air into cavities or the escape of fluids, as fœcal matter, out of mucous canals.

There are many other combinations of the splint with

the suture, as, for instance, the *Quilled Suture*, which is
used when a wound is deep and long; and there seems

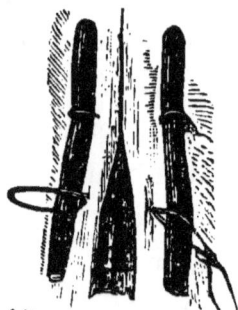

Fig. 17.

no reason why its deeper tissues
should not adhere as well as the
superficial ones (fig. 17). Then rigid
bars, called *quills*, are laid parallel
to the raw surfaces, with the double
object of resisting any drag, and
of pressing them closely together;
this manœuvre is but seldom used
in the ordinary rôle of surgical operations, except in that for ruptured perineum. Quill
is a mere *façon de parler*, and is represented by bougie,
gum-elastic-catheter, etc., strips of leather, buttons—
anything, in short, the surgical instinct suggests. Cut
them rather longer than the wound, pass a strong thread
or wire, by means of a needle in a handle, deeply
through the tissues traversing the wound. When it
appears at the opposite side, do not unthread it, but
put your finger or a forceps into the loop, and pull
the needle out; into this loop put the quill (*i.e.*,
bougie), and repeat the process as often as you would
put stitches in an ordinary wound. Take care, however, that they are not *too* numerous, that the quills are
fulfilling their office of pressing the parts together, and
yet not exerting such an amount of pressure as may inter-
cept the supply of blood to the skin edges. The quills,
of whatever substance they are made, must be smooth,
sufficiently rigid to resist considerable traction, and not
so narrow as to cut into the skin should swelling occur.

In most wounds there is some slight dressing laid over the sutures. In old times, balsamic ointments and lotions irritated the new-made wound, producing inflammation ; but, as before remarked, modern surgeons merely protect the part, and leave it undisturbed. The first dressing is of lint, wet or dry, according to the ideas of the surgeon. If the wet lint is allowed to become warm and almost dry, after a time it acts very much as a poultice, encourages an undue amount of action in the wound, and does not favour immediate union. This may be the reason why many are averse to water dressing, and prefer that their first applications should be dry. It does not seem to make any very great difference to the wound which is used so long as the covering is a protection, and leaves it alone to heal in its own way. The objections to dry dressings are, that they become dirty ; they get into close apposition with the wound, and resemble a scab ; they are difficult to remove unless first wetted, and wetting which must be continued for a long time; they become gradually soaked with the secretions of the wound, if there are any, these putrify, and a warm rotten poultice, confining the humours in the wound, and painful to touch,

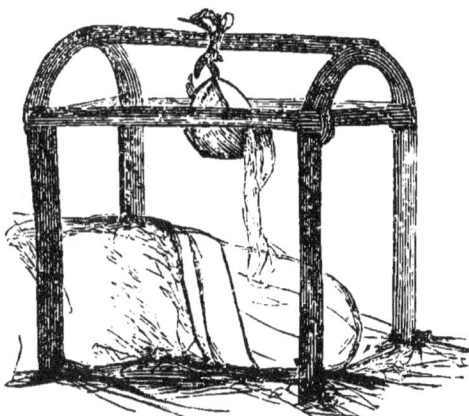

Fig. 18.

is the result. Whereas wet lint or linen, or even plain water, can always be changed; a little arrangement, such as cut No. 18, can be extemporized should the patient be unable to attend to it himself, which will ensure its being really *water* dressing.

In warm weather dressings feel very oppressive to the hyperæsthetic limb, and may often be dispensed with.* It is only too probable that the wound will not heal in its whole length by the first intention—perhaps not at all, but still the parts must be kept together by strips of plaster or wet lint. The stitches in such a case, having done their work, and being now useless, may be removed.

For the first few days, a wound of any extent which has not healed by the first intention, exudes a thin watery discharge, mixed with rusty-coloured blood, and then a rich yellow-coloured pus, like melted butter. The wounds of fat persons sometimes give exit to large quantities of oil.

In warm weather especially, a faint odour arises from the wound, disagreeable to the attendants, and sometimes to the patient himself. This may be counteracted by deodorising materials, as the chloride of zinc, or, what is better, a little peat charcoal powder laid between two layers of cotton wool, covered with muslin, and quilted like a cushion, then laid over the part. A slightly stimulant lotion is also very useful,

---

* I have treated amputation cases with no other dressing than a piece of muslin laid loosely over the stump, and occasionally a little stream of fresh air blown from a bellows.

such as a teaspoonful of tincture of myrrh in a pint of the water in which the lint is soaked.

It frequently happens that, after a few days, the wound, which previously seemed healing, rapidly comes to a stand-still—its discharge thinner and less copious, and the granulations a pale pink, like the palpebral conjunctiva. The red lotion, or some other stimulating application, may now be used ; and should the patient appear languid, perspire much at night, or refuse his food, a little quinine and diluted sulphuric acid will probably benefit him and the wound.

Should there be any unhealthy action in the wound, either secondary or primary, amounting to phagedæna or sloughing, then it is necessary in many cases to destroy the surface with nitric acid. The part should first be dried with a piece of lint, a little oil smeared over the healthy skin in the neighbourhood, and the acid applied either with the spun-glass brush of the shops, or the flattened end of a chip of wood— an old fashioned lucifer match, for instance ; this should be dipped in the acid and dabbed on the sore ; it is extremely painful, and should be done quickly and thoroughly. A bread or yeast poultice should be then laid over it till the slough separates, when water dressing may be applied.

Instruments for examining wounds.—Some wounds and sinuses, the results of disease, being too narrow and deep to be satisfactorily explored with the finger, an excellent substitute is found in the *probe.* This useful instrument is the simplest in surgery, being

merely a wire of silver or electro-plate, slightly bulbous
at one end, so as not readily to enter undivided
tissues.

Most probes have a bodkin eye at the end,
others are made with bayonet points, but the eye
probe is the safest and most popular, as it serves
the purpose of a seton needle, to draw a tape
along under the skin, and is otherwise occasionally of use. Its blunt end, too, when
bent, is the most handy instrument
for withdrawing small bodies from
narrow passages, as the external audi-
tory meatus. Some Anel's probes
are made very fine, to traverse such
canals as the lachrymal canaliculi.
For military purposes the silver probe
is made longer and
stouter, but the shape
figured is retained. For
exploring the urinary
bladder, it is generally
made of steel; its bulbous end is given a permanent bend, more or less abrupt, instead of
the eye it has a handle, and is usually called
a *sound*.

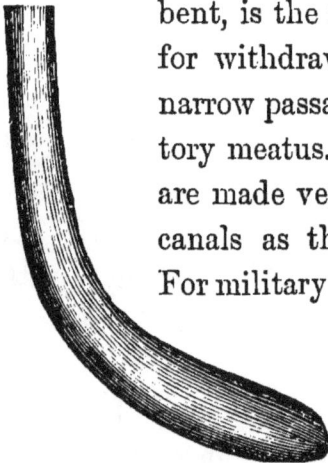

Fig. 20.

Fig. 19.

HEMORRHAGE,

In an open wound, may be seen to come from three sources, from arteries in a bright scarlet jerking stream, from the veins in a dark continuous flow, and oozing from small vessels. The relative danger of these depends on the situation and size of the vessels, and although the immediate loss of blood from an artery is perhaps greater than from its corresponding vein, the veins are so irregular in course and number, so apt to sink among the loose cellular tissue in which they lie, and so apt to resent any interference with their coats, that most surgeons dread injuring them in operations at the root of the neck, in axilla, pelvis and popliteal space, more even than arteries. The oozing varies in quantity with the patient and the tissue wounded. In punctured wounds, the aperture in the skin may not be opposite that in the vessel, and the blood, not having a free exit, may diffuse itself among the tissues of the limb so as to endanger its safety, even if the immediate danger is averted.

It is well to remember, that a small vessel wounded close to its origin from a larger, may bleed nearly as much as the parent trunk.

Treatment.—Except in very hemorrhagic individuals, the oozing generally ceases of its own accord; should it persist after the wound has been closed, the latter should be reopened and all clots removed.

Should this and exposure to the air be not enough, cold or styptics may be tried; of the latter the best, perhaps, is the tincture of the perchloride of iron, which may be kept in glycerine and applied by a little dabber of lint, the matico leaf either dry or in infusion, turpentine, Ruspini's styptic; all have much the same action. Unfortunately, those which are strong enough to be worthy of much confidence so change the surface of the wound, and cake the blood, as to interfere with the healing process; others, as the agaric, owe most of their virtue to the pressure necessary for their application.

Venous bleeding may generally be arrested by pressure, which should be made directly on the vessel; and should it appear likely to continue for any time, pads of lint or sponge may be substituted for the finger. If the vein be large, and only partially divided, the lips of the wound may be seized by forceps, and tied so as to leave a passage for the stream of blood.

Wounded arteries, even of small size, should be secured by the ligature. If in an open wound, the cut end of the vessel should be seized in forceps with blades similar to those figured, and gently drawn from the surface, so that an assistant may include it in a reef knot. The thread should be carried over the forceps point

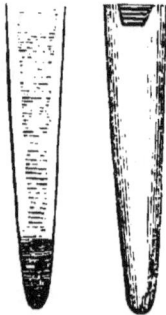
Fig. 21.

Fig. 22.

by the two forefingers, not tightened until quite down on the artery; not with a jerk, nor pulled too tightly. A very moderate amount of compression suffices, whereas violent pulling at the thread may break the latter, displace it from the vessel, or cut through the tissues, which in some subjects are very friable. I have seen, after removal of a mamma, an anterior intercostal branch break through awkward manipulation, until there seemed some danger of its retracting into the thorax. After the vessel is tied the surgeon cuts off one end of the ligature close to the knot. When an artery of any size, such as the facial or those of the forearm, is cut, it is more prudent to tie both ends even should the distal one not bleed at the time. The instrument for seizing the vessel has many modifications. Some prefer to catch it with a sharp-pointed hook or tenaculum, which, in cases where the end of the vessel is difficult to isolate from the other tissues, is very useful, and may be obtained in a clasp handle, to suit the pocket case. When the surgeon has to tie an artery by himself, he must hold the forceps in his left hand, and manage the ligature with the ring and little fingers of that hand, sliding the noose down along the forceps with the middle finger. A little practice makes this a very easy manœuvre. But it is well to have forceps which can be kept close by a slide as in Savigny's, a spring as in Assalini's, or a combination of both—the slide passing into a ring on one of the blades, as in Luer's and the torsion forceps. The latter, though very strong, requires much care to keep in order, as clotted blood or rust will in a very short time make it wholly useless. In Assalini's forceps the spring may be pushed to one side when not required. When securing vessels at the bottom of deep wounds, the ligature is very apt, in unaccustomed hands, to tie the blades of the instrument, instead of the artery. With a view to removing this difficulty, the blades have been made convex and tapering to a point, so that the noose may more easily slip over them. The blades, though generally in a line with the handle, are sometimes curved, as in those of M. Gensoul of Lyons, or with a hinge at which they may be bent, as invented by M. Colombat. The points are endlessly

Fig. 23. varied, from hooks to all degrees of roughness.

Although it is well to have such simple aids by one as a spring or slide forceps, such mechanical contrivances will often be found wanting when most required, and the student should practise his fingers until they render him nearly independent of all other prehensile weapons.

When an artery is wounded, the amount of bleeding depends to some extent upon the direction of the cut. Thus, if the latter be transverse or oblique, it will gape more than if longitudinal, and the exit of the blood is not impeded by contraction and retraction, as when the vessel is quite severed. The treatment of such an injury varies somewhat with the vessel injured, but the surgeon should never lose sight of Mr. Guthrie's two rules (p. 188)—

"1. That no operation ought to be performed on a wounded artery *unless it bleeds*.

"2. That no operation is to be done for a wounded artery, in the first instance, but at the spot injured, unless such operation not only appears to be, but *is*, impracticable."

Compression may be tried in all cases at first by graduated compresses and a bandage, or by the fingers. When the vessel is small and lies far from the surface, and attempts to reach it are likely to be injurious to other tissues, as in wounds of the palmar arches, a pyramidal pad of lint, enclosing some hard substance, as a small coin, may be placed with the apex over the bleeding point and secured by a bandage, including the fingers, should very severe pressure seem necessary. But it must never be lost sight of that the blood may insinuate itself among the deep structures, and should pressure appear inefficient, operative interference should not be delayed, as then not only will the patient be less able to bear an operation, but the performance of the latter will be more difficult, and the coats of the artery less likely to heal. Arteries such as the radial, which appear very superficial in a healthy state of the parts, are sometimes most difficult to find after interstitial bleeding.

The instruments required are—1. A Scalpel, with

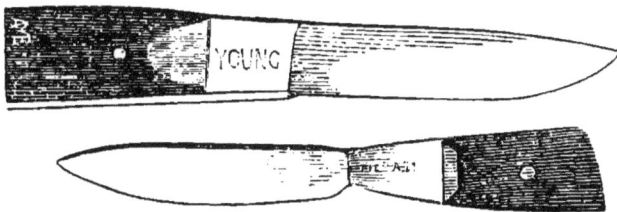

Fig. 24.

the blade shaped like the section of a finger-tip, and a

flat handle, which is very useful in gently pressing aside tissues, where a free use of the blade would be dangerous. Some surgeons have recommended silver knives for such operations, as less likely to produce accidents, but their very bluntness requires an undue amount of violence, and every practised dissector knows that the sharper his knife is the less likely is it to wound either the artery or neighbouring

Fig. 25.

tissues, if the hand be light and steady. 2. A Forceps, such as already figured, large enough to fill the hand.

3. A Blunt Hook, with a round eye near the end, large enough to let the ligature run easily through. In selecting an aneurism needle, care must be taken to see that neither the edges of the eye or the point cut, as the former may injure the ligature, and the latter prick either the arterial coats or the accompanying vein. Some prefer to have the needle placed at right angles to the handle. There should not be too much spring in the shaft, neither should the latter be so soft as to bend.

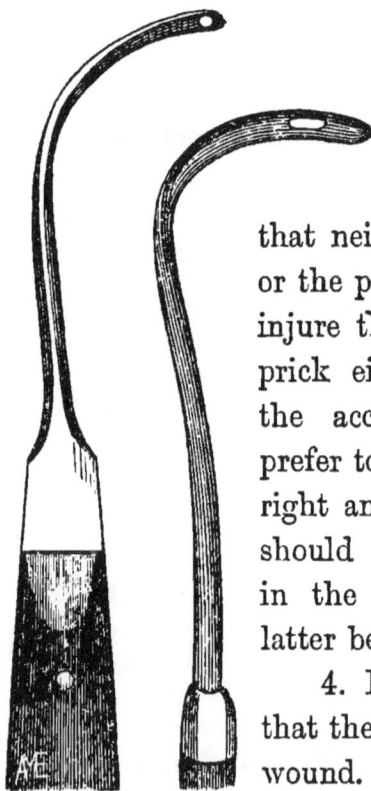

Fig. 26.

Fig. 27.

4. Ligature of such a length that the ends will hang out of the wound. The material of the thread varies with the taste of the surgeon.

Mr. Guthrie preferred dentists' silk, Mr. Fergusson prefers the strong twine used in the manufacture of Paisley shawls; but the material does not seem to be of much importance, so long as it is not too thick, and yet bulky enough for easy manipulation.

5. *Retractors* are sometimes required in deep operations ; the best are the strong blunt steel hooks, set in handles, called Syme's retractors. Bent copper spatulæ are sometimes used, but are apt to straighten under the necessary pressure.

6. Needles and thread.

In cutting down on a wounded artery, the knife should be carried in such a direction that the line of incision shall include the original wound. On reaching the vessel the needle should be cautiously passed under it, the ligature seized with forceps, and held until the needle is withdrawn, then the loop cut, and each half of the ligature tied so as to include the arterial wound. Should there have been time for a false aneurism to have formed, it is well to tie the distal end of the vessel first, as it is generally the most difficult to find, unless, of course, circumstances render immediate seizure of the upper and bleeding end necessary.

Temporary arrest of hemorrhage is required in nearly every case, and, except at the root of the neck, there are few surgical parts of the body where it is not possible to press on some proximal part of the artery. In thin persons it is possible to compress the abdominal aorta itself by relaxing the abdomen and pressing the hand firmly to the left of the umbilicus.

When obtainable, digital pressure is to be preferred, and, as before remarked, a very slight degree is necessary. Where the artery lies at such a depth from the surface that its efficient compression is doubtful, an incision may be made, into which the assistant inserts his finger. This method, scarcely alluded to in most text books, is sanctioned by the high authority of Mr. Syme, who has applied it to the internal maxillary and sub-

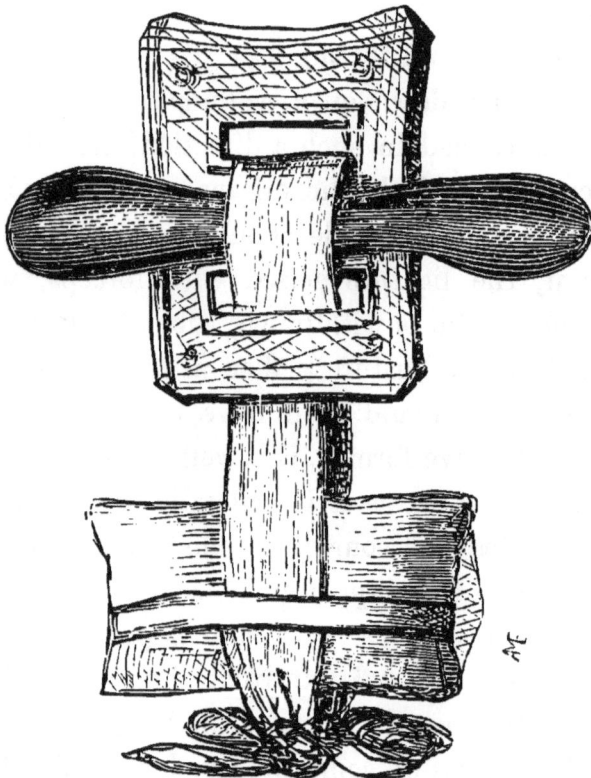

Fig. 28.

clavian arteries. The tourniquet has, however, to be substituted for the finger on many occasions.

The tourniquets in general use are all, more or less, modifications of the simple instrument devised by Morell at the siege of Besançon in 1674, consisting of a strong bandage passing through two slits in a piece of leather, a pad stuffed with hair, and a stick inserted in the loop above. Taking his idea from the method by which muleteers tightened the ropes round their bales of goods, Morell tightened his tourniquet by twisting the stick ; this had the disadvantage of compressing the whole circumference of the limb. Subsequently another pad was added to afford a counter-pressure to one placed over the artery. To enable the surgeon to regulate the pressure more accurately, and to dispense with an assistant, whose sole duty was looking after the stick, Petit in 1718 drew the bandage through two pieces of wood, which could be separated from each other by a screw. The pieces of wood have been since superseded by a brass bridge, and the instrument thus modified is that most commonly used; Luer's field tourniquet consisting of a strong steel spring and wooden pad. The spring has a slit and a row of inverted teeth at each end. When applied, a band is passed round the limb and through the slits, the pad adjusted over the artery, and the band pulled tightly and hitched on the teeth. The simplicity of this tourniquet is attractive, and may be useful in the field, but it is not so efficient as Petit's under ordinary circumstances.

In applying the screw tourniquet the following points have to be attended to :—The strap must be sufficiently strong to bear any reasonable amount of tightening. The upper bridge must be down on the lower, and the buckle not too near the bridge. Some surgeons first apply a few turns of the bandage, retaining part of the roller for a pad. Some prefer the screw being immediately over the latter, while others prefer it on the outside of the limb. When the strap and pad are adjusted, the tourniquet should be at once tightened, as a partial constriction congests the limb below, and leads to an unnecessary loss of blood.

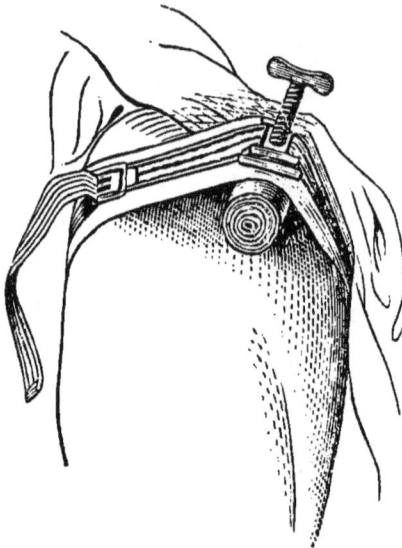

Fig. 29.

Secondary hemorrhage may occur at any time until the wound is healed, but is less to be dreaded after the ligature has separated, though it has often taken place from large arteries at a more distant period. Mr. Guthrie says,—"When a secondary does occur, it is usually from the second to the fourth week." It most commonly takes place when the ligature is about to separate. Dr. Crisp gives the following table, shewing when that event may be anticipated.

Average.				Day.
Common Iliac	.	.	.	18-25
Internal „	.	.	.	
External „	.	.	.	22

Average.			Day.
Femoral Iliac	.	.	18
Subclavian „	.	.	17
Carotid „	.	.	21
Brachial „	.	.	41

PARTICULAR ARTERIES.—Common carotid.
Guide—Anterior edge of sterno-mastoid.
In front—Twigs of superior thyroid artery and descendens noni nerve.
Behind—Sympathetic nerve.
Inside—Recurrent nerve.

In the sheath the vein is external to the artery, while between and a little behind is the vagus nerve. In some subjects the vessel is overlapped by the thyroid body, and on the left side is very close to the œsophagus.

The vertebral and inferior thyroid arteries were tied successfully by M. Maisonneuve, below the sixth cervical vertebra, for a bullet wound. The incision was along the anterior edge of the sterno-mastoid, the carotid sheath exposed and held aside.

The right subclavian has been tied in the first part of its course several times; of ten cases nine died from hemorrhage, one from pericarditis and pleurisy, nearly all presented symptoms of irritation of the pneumogastric nerve, and the most successful only survived the operation thirty-six days. In this part of its course, the right subclavian artery has the subclavian vein in front and somewhat below; the internal jugular vein on the inner side; in front the vagus, and sometimes

one or more of the cardiac nerves ; the phrenic passes
into the chest, between the great vein and the internal
mammary artery; behind is the recurrent nerve, passing
upwards to the larynx ; below is the pleura. This
complication of important structures in a space scarcely
an inch and a half in width, and the three large offsets,
the vertebral, internal mammary, and thyroid axis,
render this part of the vessel very unsuitable for any
surgical operation. The corresponding part on the left
side may be considered out of reach. When, however, the subclavian has passed behind the anterior
scalenus muscle, and is curving down over the first rib,
it may be easily reached in the triangular space bounded
by the posterior belly of the omo-hyoid above, the
clavicle below, and the posterior margin of the sterno-
mastoid in front. In this space the vessel lies at a
variable depth, and to reach it the platysma and cervical fascia must be divided. The external jugular, if
much in the way, may be tied in two places, and cut
between the ligatures. The transversalis colli artery may
be met with; and in a large majority of cases, the posterior scapular artery will be found arising from the
subclavian. In front of the latter, tucked under the
clavicle, and bound to it by fascia, are the vein
and the supra-scapular artery ; above are the nerves
of the brachial plexus, the cord formed by the eighth
cervical and first dorsal lying in close contact with
the artery, the fifth and sixth sending down, at right
angles to the latter, a small twig to the subclavius.
Although the ligature of this part of the vessel has

frequently been successful, it seems probable that, for
the future, instead of having recourse to it for the cure
of axillary aneurism, surgeons will prefer operating at
the seat of the latter, as recommended by Professor
Syme. The posterior belly of the omo-hyoid, generally
the upper boundary of the subclavian triangle, is given
in many books as a guide to the artery. It is so,

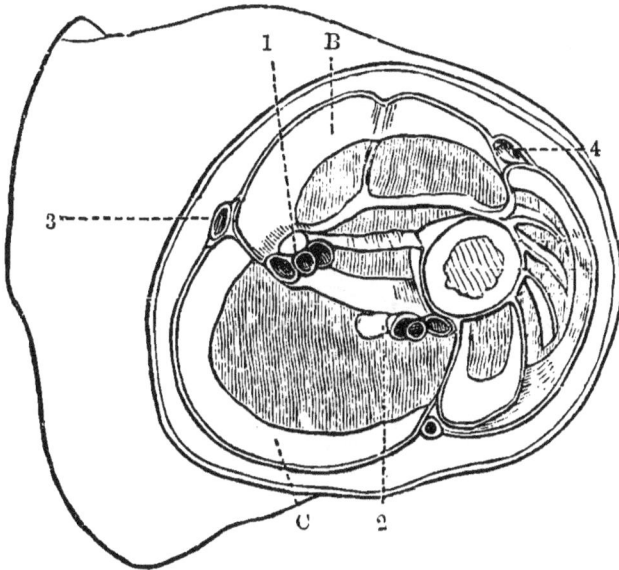

Fig. 30.
Section of the arm an inch above the insertion of the deltoid,
reduced from Bourgery.
B. Biceps muscle. C. Triceps. 1. Brachial artery, with a vein on each
side, and the median nerve above. 2. Superior profunda, venæ comites,
and musculo-spiral nerve. 3. Basilic vein. 4. Cephalic vein.

inasmuch as the vessel lies on a lower level, but frequently it is in front of the artery, obliterating the
triangle, owing to an expanded origin from the clavicle.
Mr. Turner found this condition in 13 of 373 subjects
in the Edinburgh dissecting-room.

Brachial—frequently tied for bleeding from wounds of the hand and forearm.

Guide, inner margin of biceps. The first incision exposes the fascia, through which is seen the white line of the median nerve; the next divides the fascia, and on drawing aside the nerve, the artery is seen with a vein on each side of it.

Radial.—Guide, inner margin of supinator radii longus; it has two companion veins, and the radial nerve is on its radial side.

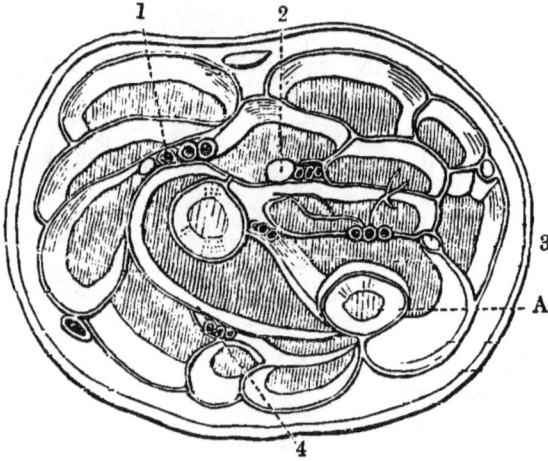

Fig. 31.

Section of the upper part of the forearm.

A. Radius. 1. Ulnar artery, venæ comites, and nerve. 2. Median nerve and vessels. 3. Radial artery, and near the ulna are seen the anterior interosseous vessels. 4. Posterior interosseous.

Ulnar is easily found below the middle third of the forearm; here it passes from under the superficial layer of muscles, and is joined by the ulnar nerve.

Guide, radial margin of flexor carpi ulnaris.

Palmar arches.—It is sometimes difficult to determine which of these anastomoses is the one wounded. But the lines on the palm afford some guide, that passing from the middle of the wrist to midway between the thumb and index finger giving an approximation to the course of the deep or radial arch. It is quite impossible to lay down rules for the treatment of wounds in this situation; and the principles with which every surgeon is supposed to be acquainted can alone suggest the proper treatment for these embarrassing cases.

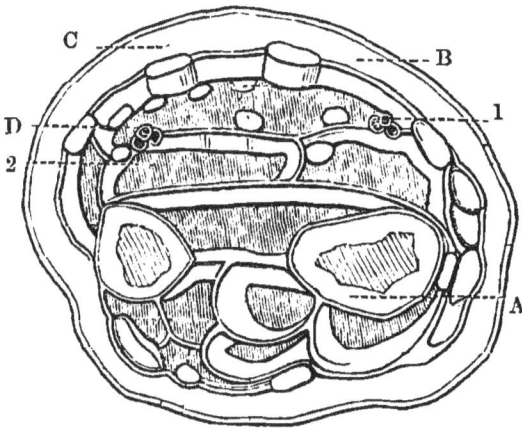

Fig. 32.

Section of forearm above the pronator quadratus.

A. Radius. B. Flexor carpi radialis. 1. Radial artery. 2. Ulnar artery and nerve. C. Palmaris longus. D. Flexor carpi ulnaris.

VESSELS OF THE TRUNK.

Intercostal arteries, "although," says Mr. Guthrie, "often injured, rarely give rise to hemorrhage, so as to require a special surgical operation for its suppression; but whenever it does so happen, the wound should be enlarged so as as to shew the bleeding orifice, which should be secured by one ligature if distinctly open, and by two if the vessel should only be partially divided." This great authority adds that twisting or bruising the vessel is frequently sufficient, and that he has tied intercostal arteries without much difficulty, except when the parts were unsound and friable. Styptics, he says, are liable to drop into the cavity of the chest.

The lower end of the abdominal aorta, the common external and internal iliac arteries, may all be reached with ease through an incision curving from Poupart's ligament towards the last rib, dividing the external and internal oblique and transversalis muscles, the transversalis facia, and then drawing up the peritoneum. The umbilicus indicates the bifurcation of the aorta, and the surgeon must not, except in very fortunate circumstances, expect to *see* the higher vessels. In front of the external iliac there is often a lymphatic gland which may give some trouble.

Superficial Femoral.—Guide, inner edge of sartorius muscle if the artery is to be tied in Scarpa's triangle, the outer if in Hunter's canal. The vein lies in the former situation, internal in the latter, behind the artery. The long saphenous nerve is in close relation with the vessel, generally outside or in front of it.

The tibial and fibular arteries are now only tied, except in very exceptional cases, for wounds, and the old rules for reaching them by incisions planned so as to avoid the division of muscle are not likely to be strictly adhered to for the future. When they are injured, the wound and the flow of blood are the best and most direct guides. Wounds of the plantar arteries are as

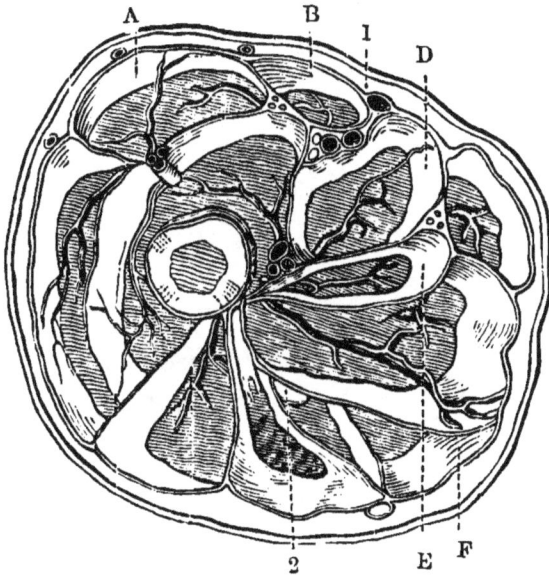

Fig. 33.

Section of the thigh.

A. Rectus. B. Sartorius. 1. Femoral artery, vein, the saphenous nerve, and that to the vastus internus. 2. Deep femoral vessels. D. Adductor longus. E. Adductor brevis. F. Adductor magnus.

embarrassing as those of the hand. I remember the case of a butcher's boy who dropped his knife between his first and second metacarpal bones. The question of ligature of the divided vessel was discussed ; but ultimately pressure was applied by means of compresses and a tight bandage. The blood did not flow externally again, but it did into the deeper tissues; and in three days the lad was dead from gangrene of the leg and thigh.

Within the last few years a new method of dealing

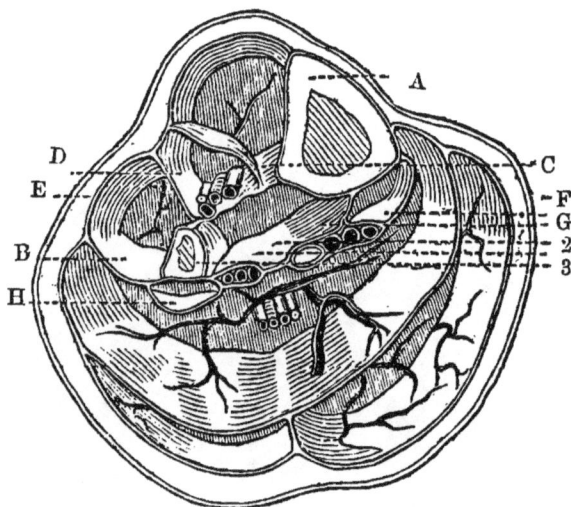

Fig. 34.

Section in the middle of the leg.

A. Tibia. B. Fibula. C. Tibialis anticus. D. Extensor communis digitorum. E. Peroneus longus. F. Flexor longus digitorum. G. Tibialis posticus. H. Flexor longus policis. I. Soleus. J. J. Gastrocnemius. 1. Anterior tibial artery, veins, and nerve. 2. Posterior tibial artery and veins, with the nerve on fibular side. 3. Fibular artery.

with arterial hemorrhage has been suggested by Professor Simpson. He argues, that for every artery tied there is a gangrenous piece of tissue enclosed in the thread, that the latter soaks up the animal juices, they putrefy, and consequently every ligature is a line of poisonous material with a sphacelus at one end ; and, moreover, that but a slight amount of pressure and short length of time are required to close the end of a divided artery, but that with the ligature you cannot allow the patient any benefit from these facts, as once the

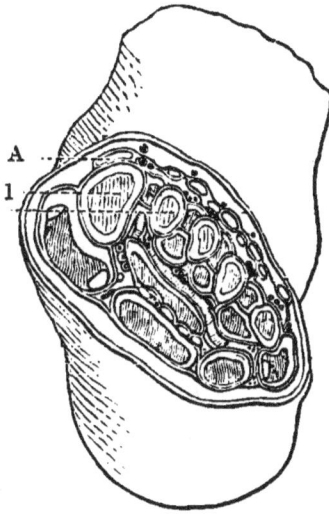

Fig. 35.

Sections through the metatarsal bones.

A. Extensor proprius pollicis. 1, 2. First and second metatarsal bones ; between the artery is seen descending to join the plantar arch.

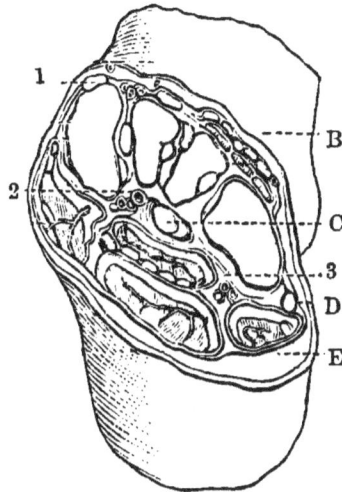

Fig. 36.

Section of the foot at the tarso-metatarsal joint.

A. Extensor proprius pollicis. B. Four tendons of extensor communis. C. Peroneus longus tendon. D. Peroneus brevis. E. Adductor minimi digiti. 1. Dorsalis pedis artery. 2. Internal plantar vessels and nerves. 3. External plantar.

ligature is tied round a vessel you cannot remove it until it is thrown off by ulceration and suppuration. Therefore he suggested that the vessels be temporarily compressed by needles passed at right angles to their course, and so far, trials of this method have been very satisfactory. Acupressure, as it is called, may be applied by means of long pins passed from without across the bleeding vessel, or by short needles from within, thus— Take an ordinary sewing needle threaded with wire, pass it under the vessel, then take a doubled wire, pass the loop over the needle's point, hitch it round the eye end, and let the remainder hang out of the wound with the wire which is attached to the needle. Between the wire on its superficial, and the needle on its deep aspect, the artery is firmly compressed, and when, at the end of a period varying from thirty-six to fifty-six hours, the needle is withdrawn, the wire is freed also, and these foreign bodies removed from the wound.

Without disrespect to the ligature, this new method may be looked upon as likely in many cases to be very efficient, and remove some of the dangers and delays of surgical practice.

With it may be classed the transfixion of bleeding points as leech bites, by needles, and twisting thread round them as in the hare-lip suture.

FRACTURES.

THE CLAVICLE is generally broken rather external to its middle. *Causes, muscular* action and more commonly direct violence. *Displacement.*—Sternal fragment is lifted by the sterno-mastoid ; the acromial portion falls, partly from the weight of the arm, partly

from the action of the muscles attached to the lips
of the bicipital groove. In oblique fractures there is
generally overlapping of the fragments; when the break
is transverse or near either extremity, the displacement
is trifling.

Treatment.—Raise the outer fragment by bending
the arm at the elbow, and pressing it upwards, draw the
shoulder back so as to bring the overlapping fragments
into their normal position. Keeping them thus, however, is very difficult, and often impossible ; the most
common method is to place pads in the axillæ, that
on the injured side being the larger. A bandage is
then passed alternately round each shoulder, as in the
figure of 8, and the injured shoulder supported by
placing the fore-arm in a sling, drawing it somewhat
across the chest, and by some circular turns of a bandage fixing it to the latter.

There have been numberless ingenious contrivances
for preventing deformity after this fracture, but the indications already mentioned are followed by all. Some
surgeons, as Velpeau, discard the axillary pad, and use a
dextrine bandage; others use straps. I have seen a
case treated with almost perfect success by Jobert de
Lamballe, by simply keeping the forearm flexed across
the chest. When axillary pads are used, it is almost
needless to warn anatomists that they may be produc-
tive of some harm by pressure on the axillary vessels.
The apparatus, whatever it is, should not be moved
before the fifteenth day ; by the twentieth the union
will probably be pretty firm.

SCAPULA.—Body and processes.

Fractures of the *body* generally result from direct violence. By tracing the spine and costæ, rendering the latter prominent by moving the arm either over the chest, or backwards to the spine. Such injuries may generally be discovered unless excessive swelling has come on.

Treatment.—Endeavour to relax the muscles by lifting the elbow, which should also be placed slightly backwards, to relax the teres-major muscle. The parts should now be steadied by a bandage round the trunk.

The surgical neck of the scapula implies that portion of the bone in front of a line drawn through the suprascapular notch.

Symptoms.—Slight depression under the acromion, hard swelling in axilla; pressure directed upwards restores the shape of the shoulder, and crepitus may then be felt. Coracoid process moves with the humerus.

Treatment.—Support the broken portion and head of humerus by a pad in the axilla. Support the arm in a sling, and bandage it to the chest.

THE ACROMION process is broken by direct violence. The arm drops, and cannot be elevated from the body, owing to the central portion of the deltoid having lost its fixed point. *Treatment.*—Supporting the arm by a sling, and keeping the elbow from the side. The re-union is frequently ligamentous.

THE HUMERUS may, by direct violence, be broken through the anatomical neck, or the tubercles, but generally gives way at the anatomical neck. In the two former, the mobility of the arm is not much altered; there is slight depression under the acromion, a fulness under the coracoid process, and the arm appears broader than natural.

In fracture through the surgical neck, the lower fragment is apt to be drawn inwards, and is difficult to retain in apposition. So the bone reunites with a permanent bend, as in the accompanying outline.

Fig. 37.

The treatment must, of course, depend on the tendency to this displacement; in some cases it is only necessary to lay the arm on a pillow with a pad in the axilla; in others to apply a dextrine bandage, but in general it is more prudent to apply rigid splints on each side, the upper end of the inner one being well padded. Splints made of thin leather, spread upon a thin layer of soft light wood, which is afterwards cut into longitudinal strips so as to form a rigid splint in its long axis and a flexible one in the transverse, are very convenient. They may be fixed by straps of plaister and covered by a bandage from the elbow upwards, reversing towards the shoulder, terminating by a couple of turns round the opposite axilla, and pinned on the injured shoulder.

The forearm should now be supported in a sling, adjusted so as not to press up the elbow, as that might tilt the lower fragment forwards and inwards.

The accompanying cut shews a very convenient

form of arm splint. It is made of thin japanned metal, has at its upper extremity a padded crutch turning on

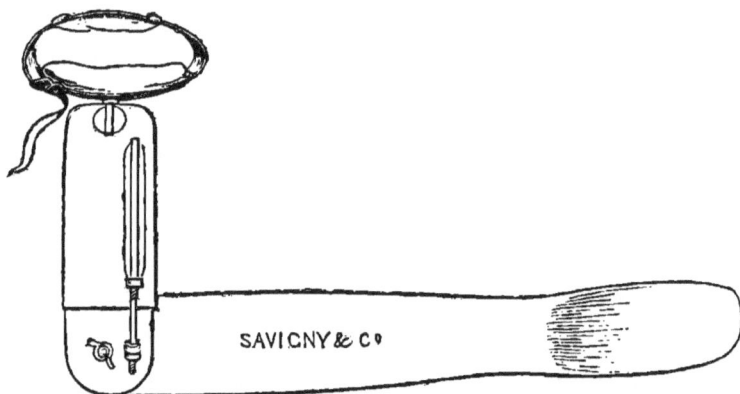

Fig. 38.

a centre pin. The distance between this crutch and the elbow may be varied by the screw near the hinge ; the latter permits the front segment of the splint to come quite round, so that the apparatus serves equally well for the right and left arms. Of course the pressure in the axilla must be carefully regulated, and may often be dispensed with altogether. The humerus frequently breaks below the middle third, the fissure is generally oblique and difficult to adjust, and the soft parts are apt to intervene between the fragments. The direction of the displacement varies as the fissure is downwards and forwards, or downwards and backwards ; in both cases the triceps displaces the lower fragment, drawing it upwards.

Treatment.—After the fragments have been carefully adjusted by gentle extension from the elbow, a rectangular splint should be placed on the inner side and secured to the arm and forearm by straps. A bandage should be applied from the hand upwards, and motion further prevented by a sling. Concave leather arm-rests are sold by the instrument makers for a few shillings. It too often happens, in spite of all our care, that the fragments reunite with some change of position, as in the cut ; and, consequently, some shortening. At other times union is delayed, or does not take place at all. This subject is fully treated of in all surgical text books.

When the humerus gives way just above the condyles the lower fragment is carried backwards so as to simulate a dislocation. If seen at an early period, the two injuries are easily distinguished by fixing the humerus and drawing gently upon the forearm. If it be a fracture, the joint will regain its usual outline and crepitus is felt, the distortion reappearing

Fig. 39.

when the extension is withdrawn, though perhaps to a less extent. It will also be observed that the condyles of the humerus have the same relation to the other bones as on the uninjured arm. The rectangular splints, or gutta percha moulds, are also used for this injury, and the patient must be prepared for some subsequent stiffness, which will, however, disappear in time. To prevent this as far as possible it is well to remove the rigid apparatus in a week or ten days, and use gentle passive motion from time to time.

THE OLECRANON may be broken by muscular action or direct violence. The injury is easily recognised by the separation of the fragment, which is drawn up by the triceps. There is usually much swelling and ecchymosis ; in many cases the reunion is ligamentous.

Treatment.—A straight splint, reaching a few inches above and below the joint, rather broader than the limb when straightened, and well padded. Laying this in front of the elbow joint, fix it by a strap of plaister at either end. Take another strap, and applying its middle to the back of the arm above the upper fragment, draw it gently downwards, and when the parts are in apposition cross its ends over the splint. As a precautionary measure it is well to apply a bandage from the hand. The apparatus should be kept on a month and then passive motion practised.

THE RADIUS AND ULNA may give way in any part of their length. When both break, the fracture is generally below the middle. *Treatment.*—Take two splints, one of which reaches from the elbow to the knuckles, and is broad enough to allow the arm to lie comfortably upon it in the supine position. While an assistant makes steady extension, fix it by straps at each end, leaving the thumb free. Now, place the shorter splint in front and bandage as far as the elbow. The bandage is not absolutely necessary, but keeps the whole compactly together and presents a neat appearance. The splints may be removed in general within four weeks. It is said that the ulna unites sooner than the radius. When the latter bone is broken above the insertion of the pronator radiiteres, the forearm presents a very characteristic appearance from the upper fragment being supinated and drawn forward by the biceps, while the lower is pronated and drawn towards the ulna by the pronators. Here two splints are also required.

This tendency to displacement of the lower fragment is greater when the fracture is at the lower third, and may be counteracted by using the hand as a lever, placing the arm in a state between pronation and supination, and laying it on a splint, the distal end or handpiece of which curves towards the ulnar side.

FRACTURES OF THE THIGH.—Of the neck of the femur.

Fig. 40.   Fig. 41.
Fracture within capsule.   Fracture outside capsule

Symptoms.—Shortening from half an inch to an inch. Eversion of the limb; flattening of the hip. Crepitus when the injured leg is drawn to the same length as its fellow. Age, above fifty.

Treatment.—Long splint of Desault attached to the

s

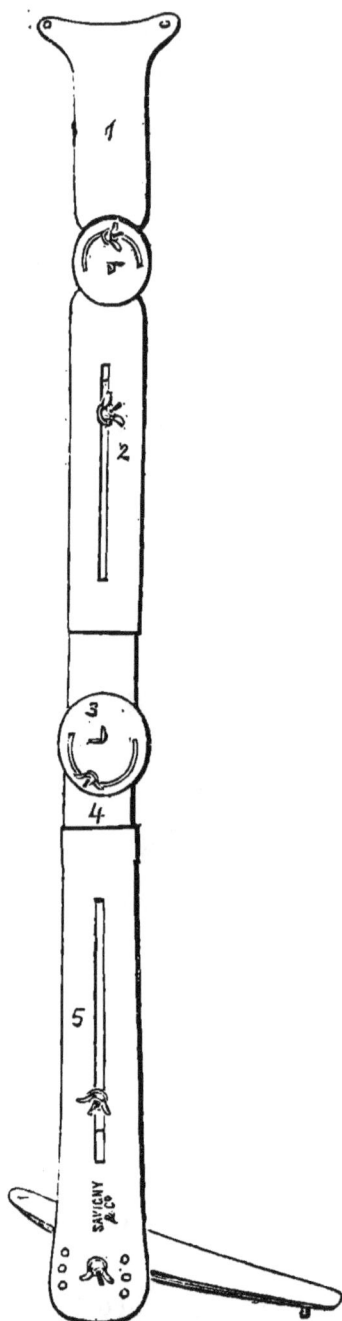

Fig. 42.

foot by a figure of 8 bandage, or straps, as previously described. A soft handkerchief, passed under the hip, is threaded through holes in the upper end of the splint, and when the limb is drawn downwards to an equal length with its fellow, the ends of the perineal band are tied so as to keep up extension. A broad bandage is then passed round the trunk. This long splint is applicable to all fractures of the thigh, and most modern authorities prefer it to the more complicated apparatus. When obtainable, however, such as is here figured, will be found of great assistance, as its length can be adapted to different cases. The foot piece is a very important addition, as it prevents to a great degree the eversion of the foot, which is so apt to occur in most fractures of the thigh. When the seat of injury is in the upper third, every anatomist will see that some additional means are required for counteracting the psoas and iliacus muscles. The most efficient are two short Gooch's, or pasteboard splints, one on each side of the thigh opposite the fracture, fastened by a looped bandage or straps.

When the femur is broken close above the condyles, the lower fragment is drawn backwards by the gastroc-nemius, and upwards by the hamstring muscles, should the fissure be in a direction downwards and forwards. Serious injuries to the structures in the popliteal space have resulted from the sharp end of the lower fragment

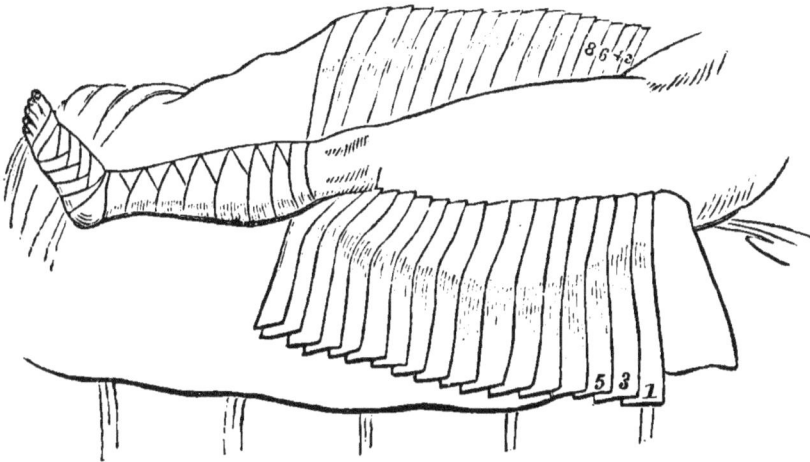

Fig. 43.
Many-tailed bandage (Scultetus) after Curling.

being drawn in among them, it is therefore advisable either to add a well padded back splint to the long one just mentioned, or to keep the knee slightly flexed on a Macintyre splint, in what is technically termed a double inclined plane.

The Patella, when broken transversely, may be placed on a straight splint or a Macintyre comfortably padded ; a bandage should then be passed a few times below the knee and pinned, then passed above the fragment which is drawn up by the rectus, and hitched over it, so as to draw it downwards. Mr. Wood of

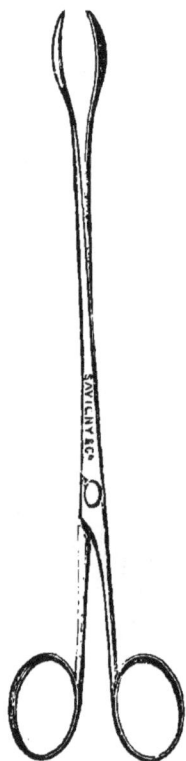 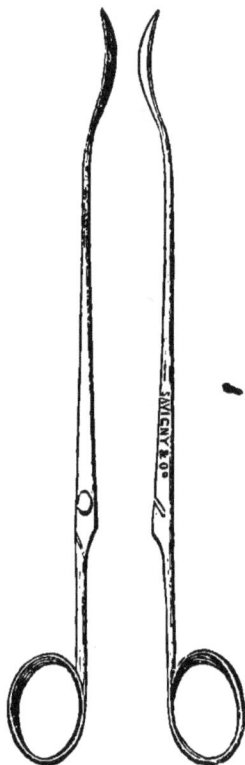

Fig. 44.   Fig. 45.

King's College, London, has fixed two hooks on the under part of the splint, round which he turns the bandage, but the screw of the Macintyre splint will be found nearly as convenient.

Fractures of the os calcis may be caused by muscular action, and must be treated as if the tendo-achillis had been ruptured. The apparatus consists of a slipper with a strap fixed to its heel, and another strap round the thigh above the knee, to which it can be attached, and by which the toes can be pointed.

Cut 44 represents the forceps invented by Messrs. Whicker and Blaise. The blades of this instrument, as seen in cut 45, may be unlocked, and either of them used singly; or they may be separately introduced into the gun-shot, or punctured wound. Thus, a foreign body or splinter may be enveloped without any stretching of the surrounding tissues; and by means of the peculiar joint, which is easily shipped or unshipped, either the firmest traction may be exerted, or the grip be readjusted with facility.

NOTE.—The field (thigh) splint at p. 258 is also the invention of Messrs. Whicker and Blaise. It is so constructed as to afford any length and any angle that may be required. The segments marked 2 and 5 respectively can be made to slide from or towards the centre 3. In case of need the upper half gives an arm-splint, leaving the lower available for any purpose requiring a short straight splint.

INDEX.

266 INDEX.

PRINTED BY R. AND R. CLARK, EDINBURGH.

In two vols., demy 8vo, pp. 1676, Price 36s.

THE OBSTETRIC MEMOIRS AND CONTRIBUTIONS,

INCLUDING THOSE ON ANÆSTHESIA,

BY J. Y. SIMPSON, M.D.,

PROFESSOR OF MIDWIFERY IN THE UNIVERSITY OF EDINBURGH.

EDITED

By W. O. PRIESTLEY, M.D.,

AND

H. R. STORER, M.D.

" *Will be consulted by all who are interested in the progress and advancement of Obstetric Science.*"—BRITISH AND FOREIGN MEDICAL CHIRURGICAL REVIEW.

" *The author of these Memoirs takes rank as a* chef d'école"— LANCET.

EDINBURGH: ADAM AND CHARLES BLACK.

Works by Sir John F. W. Herschel, Bart., K.H., &c.

I.

In crown 8vo, cloth, Price 7s. 6d.

PHYSICAL GEOGRAPHY.

" *An admirable manual of the whole science.*"—BRITISH QUAR-TERLY REVIEW.

" *The Book is a most fascinating one.*"—EDUCATIONAL TIMES.

" *It is utterly impossible to give an account of the immense amount of information so admirably and lucidly compressed in the Volume before us.*"—LONDON REVIEW.

II.

In fcap. 8vo, cloth, Price 5s.

METEOROLOGY.

" *Contains a brief but elaborate survey of the whole domain of Meteorological Science.*"—BRITISH QUARTERLY REVIEW.

" *As Text-Books for College and School use, on the subjects on which they respectively treat, there is nothing in the whole range of our educational literature, which can at all be compared with them.*"—EDUCATIONAL TIMES.

III.

In fcap. 8vo, cloth, Price 3s. 6d.

THE TELESCOPE.

EDINBURGH: ADAM AND CHARLES BLACK.

www.ingramcontent.com/pod-product-compliance
Lightning Source LLC
Chambersburg PA
CBHW021515210326
41599CB00012B/1264

* 9 7 8 3 3 3 7 1 9 1 8 2 5 *